ALL CAB DRIVERS LOOK ALIKE

ALL CAB DRIVERS LOOK ALIKE

THE SCIENCE OF CHANGING PERCEPTION THROUGH EXPERIENCE

FELICE L. BEDFORD

TIVOLI PRESS

Published by Tivoli Press
USA ITALY
Philadelphia, Pennsylvania
Manufactured in the United States of America

Prism on cover adapted from image by Lexi Sinnott, licensed from Shutterstock Images LTC

ISBN-10: 0615734715
ISBN-13: 978-0615734712

Library of Congress Control Number: 2012955222

For my mother, Pauline Bedford nee Allalouf.
She gave me life, inspiration, confidence, genes,
perspective, empathy. They are all for you, but
here is the one I promised. One day—soon—we
will be together again.

Contents

Preface

This book was written with my grandmother in mind. Her name was Flora Allalouf. And I thought of your grandmother, too. Of course, it is also for my mother, to whom it is dedicated, and all parents. The book is designed for curious people whether they have received a formal education or not. It is also for my students, past, present, and hopefully, after they read this work, future. My colleagues will discover much that is new, if they can tolerate a style they have been brainwashed to dislike—one that talks to the reader and tries to use ordinary language. I wish I could say something like the first 25% of each chapter is well-known in the field and the remainder is new. This way, skeptical colleagues and hurried readers could skip to the interesting bits. But, unfortunately, the book is not like that. I will say that there is a lot of new thought here, even when I don't announce it as such. The book is also for Alan Alda, the actor (who must be interested in science books), and Sir Tom Jones, the singer (if he reads books). Importantly, it is for me, perhaps selfishly or irrationally. Because I will forget? To move on? For posterity? I'm not sure.

I put the answer to the question first. The question is: Who is this book *for*. I can see my critics saying: "Bedford one moment explains and re-explains things in words a child could understand and in the next, drops a complex concept or equation (!) without elaboration." All true. Space and time are limited and this is how I can offer information for all of us.

It is for Alan Alda, by the way, because he says about science: "To often, I think, we hear about the results of the search without the drama of the search itself...Invite us to solve the puzzle along with you, with all its emotional ups and downs...They make what you do human to us." [1] This is something I believe in. I love the process of intellectual mystery solving and believe others will love it, too, if they can experience it that way. But I was reading Alda's book in the first place because I already always wanted to meet him. Explanations for motives are never simple. If he wasn't a talented actor, director, and science advocate, I'm sure he would have been a great colleague. I better not explain why the book is for Sir Tom Jones.

In the same spirit of the science process, this is not a book about perceptual learning. It is a book that *I wrote* about perceptual learning. Separating the "facts" from the theory and the conclusions from the scientist can never be completely achieved. I'm sure more people would love science if fewer people portrayed science in a distilled "just the facts, ma'm" [2] manner. Psychologists are a particularly defensive bunch when it comes to this issue and there are still some stodgy journals that require every single instance of the word "I" in a submitted manuscript be stricken before publication, as if the experiments got up and ran themselves.

What this book is about is the things you take for granted when "truckin' on down the avenue" [3]. A quarter of a century ago, I set out to understand all the ways that experience alters our perception. Many of you may be disappointed though, because I do not mean perception in the sense of how you feel, or what your opinion is, about the people and things around you. Instead, I refer to what you literally see, hear and so on. Like I said, the things you take for granted. Space, time, growth to list a few, but also perceiving faces from other races, cell phones that feel like they are part of the body, and becoming an expert at the taste of wines.

There are so many examples of perception becoming altered through experience, but they were all scattered in different literatures and even in different eras. In addition, at least one big one has never before been recognized as a perception change. I discuss this one towards the end of the book in Chapter 11. Even more frustrating, there was no unifying framework for understanding all the changes and how they relate to one another and to the rest of learning. I believe I have succeeded at the goals and present the discoveries here. I also draw heavily from both my own research and teaching, as will become apparent. Research and teaching became more intertwined in my quarter-century quest than I ever thought possible.

The first chapter introduces the area by trying to hit people over the head with it—as literally as one can in black and white characters. I discuss how disorders that can result from being hit over the head can change perception. Perceptual learning and perceptual disorders have not been related to one another before. The three chapters after disorders introduce problems to get people to start to think about how normal perception is changeable, not stable. I do this by discussing what happens when we grow (Chapter 2), use tools so much that they become parts of ourselves (Chapter 3), and the two types of errors people make in sports that correspond perfectly to two major types of changes in perception (Chapter 4). The next three chapters (Chapter 5, 6, and 7) are about one of the major types of perceptual learning, which I refer to as perceptual expertise. It is also known in the scientific literature as *perceptual discrimination learning, differentiation*, and, unfortunately, *perceptual learning*, an ambiguous term which leads to confusion because it is so general. It is through this process of perceptual expertise that we can learn to be experts at so many things—the taste of wine, X-rays, sheep faces—and what gives us trouble recognizing people from other races. These chapters also contain the inspiration for the title of the book. The three chapters after that, following a prelude on your senses at war (Chapter 8), are about the other major type of perceptual learning, that of perceptual adaptation. These are on space (Chapter 9), time (Chapter 10), and—believe it or not—the seemingly mysterious effects of meditation and guided imagery therapy on health (Chapter 11). I argue that the immune system is a perceptual sense modality like vision or hearing and its adaptation can lead to healing. Chapter 12 discusses some of my overarching theory not already covered, as well as tying up loose ends and forecasting the future. Finally, Chapter 13 is an aftermath of activities to apply the abstractions to actually improving perception intentionally.

Two of the chapters are especially for students in the sense that they are so detailed that only those under threat of a grade book will likely read them. One other chapter will likely appeal to my colleagues—probably the wrong chapter—because the abstractions found there speak more in their language. Two or three further chapters are my favorites...Ok, enough analyses.

I don't know if I "told it right" (more Alda) but it was exciting. Enjoy the journey.

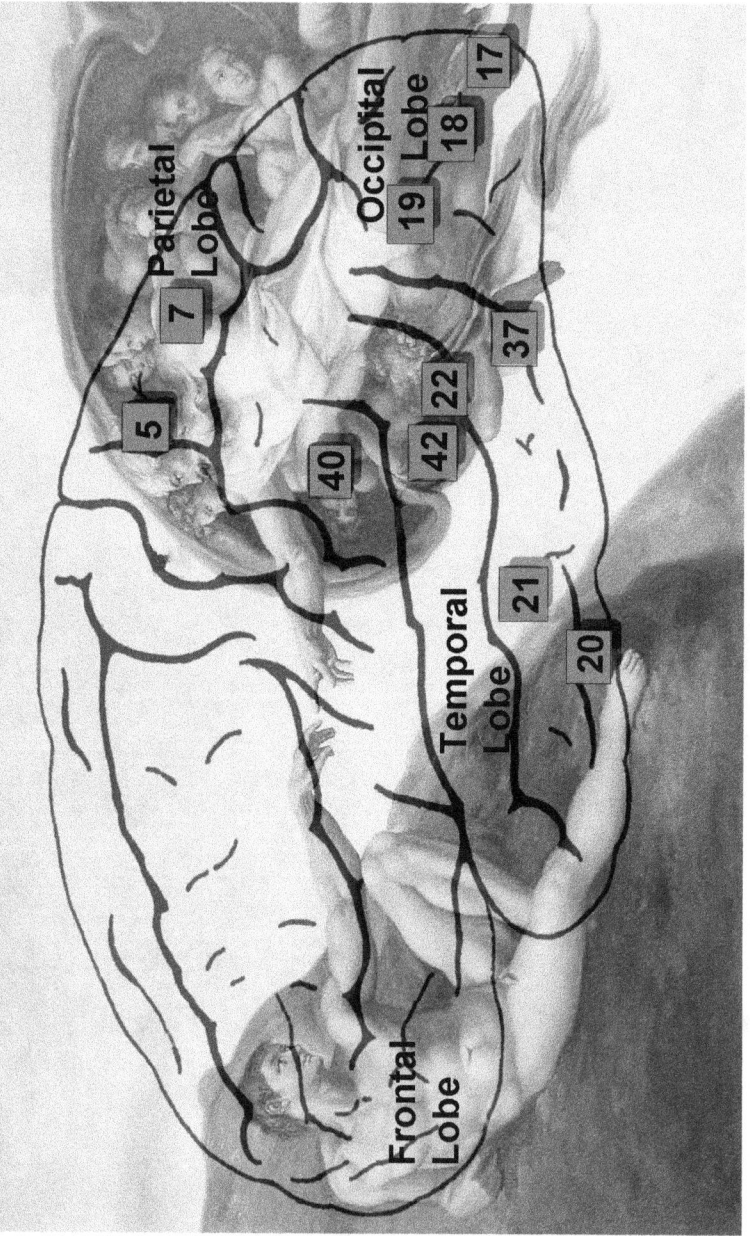

Figure 1-1 Location of areas in the brain leading to perceptual problems. Shown in the background is The Creation of Adam. The images of God and the angels (right) were intentionally painted in the shape of a brain but it took 475 years to uncover Michelangelo's secret.** *Foreground: adapted from 20th ed. Grey's Anatomy.* Composite by Felice Bedford and Laura P. Garcia

When Perception Goes Very Very Wrong!

Can you pretend to be a patient? Imagine that twenty-eight slips of folded paper are making their way around to you in a baseball cap. If you decide to pick one of the papers out of the hat, your job is to look up the disorder named on the paper and act as if you have the disorder. If you do not want a disorder then you can be a "therapist" who tries to figure out what is wrong.

Select a paper. It says "hemineglect". Investigation on the web would quickly reveal a deficit where half of the world is ignored. A patient might act as if everything on the left side did not exist. He would not wave hello to anyone on the left, would not pick up a cup of coffee if it were on the left, and may bump into obstacles—but only on the left. It is not as if he is just partially blind. A blind person would turn his head or body to see things on the side where he could not see. Instead, the *hemineglect* patient behaves as if half of space has simply vanished! Information can get through the eyes, but is ignored by the mind. He might even stop combing the hair on the left side of his head or shaving the left side of his face. If asked to draw a picture of a cactus, he might draw half a cactus. A popular neurological test provides a pre-drawn circle and asks patients to fill in the face of a clock. Hemineglect patients may squeeze all twelve numbers of a clock face into just the right half of the circle.

If you were pretending to have hemineglect, you could act out all of these behaviors. When I introduce this role playing game in my class, I remind everyone that they cannot just tell the therapist volunteers the name of their specific brain disorder. As a real patient, there would be no way to self-diagnose hemineglect any more than hemochromatosis (too much iron in blood). Instead, they should play act as if they had the disorder or explain symptoms if they seem bothersome—just like at a physician's office. Likewise, a clinician would not be able to just ask the patient for the name and definition of his disorder if the patient had never before been diagnosed. Those with disorders are asked to raise their hands (if their disorder allows them to raise their hands...); they stay seated while therapists circulate, question, and take notes.

Some "patients" are astonishingly good at keeping in character. It's trickier than it seems. It is very hard to resist the temptation to give hints when the questioner is struggling while it is so obvious to you. It is also difficult to think of precisely what you would complain about or how you would behave on a minute-to-minute basis if you really had the disorder. Other people seem to instinctively know what to ask the patients in order to hone in on the true nature of a disorder. This, too, can be a challenge. Try one as a therapist:

You: Um, what's bothering you?

Patient: I can't see!

You: (Buying time to think) You...can't...see?

Patient: I was in a motorcycle crash and hit my head. Before that I was fine.

You: Can you see anything?

Patient: I can see things over there (points right), but not on this side (points left) unless I turn to that side.

You: (Hey—is this hemineglect that I just learned about? But hemineglect patients don't turn to the blind side; they don't even know it exists. The left side of his hair and face seem just as groomed as the right. But just to be sure...) Can you draw a clock for me on this piece of paper?

Patient: Sure. (He holds paper on his right side, but draws clock fine—it is a whole clock and all the numbers are where they should be.)

You: (Doesn't look like hemineglect. Could he just be blind on

one side? Could it be that simple—and dull?) Was your disorder just partial blindness? Sometimes accidents leave
people blind for only some regions of space.

Patient: No. I've been to a doctor. She said my eyes were fine and
also there's nothing to indicate that I should be blind.
And…well, I'm not supposed to give you any hints. (Sigh.)

You: (Sigh.) Do you see this? What is it?

Patient: Yes, it's a pen

You: (Holds up same pen on his left side.) Do you see this? What
is it?

Patient: No. I don't see anything.

You: (You try waving the pen around.) How about now?

Patient: No.

You: (You notice his eyes are following the pen as it moves!) Are
you sure you don't see anything?

Patient: Positive!

You: What direction did I just move it?

Patient: Up

You: (Correct!) How about now?

Patient: Left

You: (Correct!) So you can see it now?

Patient: No. I told you I can't see.

You: If you can't see the pen, how did you know it moved up and
then to the left?

Patient: It just seemed like there was motion upwards; I don't
know. I guess I just guessed.

You: (You quietly put your can of Pepsi—it's a Pepsi campus—on
his left. He says he sees nothing. You ask him to guess and
pick up the object anyway. He reaches out, and you notice he
does so in roughly the right direction. What's more, his hand
seems to form into the right size and shape to hold a Pepsi
can—not a pen, or a knapsack, but a soda can.)

You: (Throws pen at subject's left side; was that frustration or
clever diagnostic?)

Patient: (Dodges pen and it mostly misses.)
You: I know! You're faking! Was that what was written on your
   slip?
Patient: NO, I AM NOT. And no, faking is not my disorder.
You: Ok. Can you walk around?
Patient: Ok. (Walks around pretty well; does not smash into
   anything.) If I were faking, don't you think I'd bump into a
   few walls just to make it look authentic?
You: Good point...

   So you seem to have someone who insists he cannot see. He does
not seem to be faking, yet behaves in ways that suggests visual information about the world is getting through on the alleged blind side.
He has enough information so as to position his hand correctly, not
bump into things, know the direction of motion, and distinguish
simple shapes if you were to test him further. On his slip of paper it
says "blindsight". *Blindsight* is the name given this disorder in which
being conscious of seeing seems to be dissociated from actually
seeing. You can "see", but are not *aware* you can see. It is a fascinating
disorder of great interest to those who study consciousness, as well as
to those who study vision.
   Would you have thought to ask him to reach for something?
Would you have noticed that his grasping was appropriate? Around
now, most students ask for my "cheat sheet". That is, they see I am
glancing at a written table that I prepared of 14 disorders (2 repetitions of each disorder had been passed around). I hear a lot of grum-
bling about how can they know what someone has if they don't know
what the disorders are. I will come back to that. First try one more.

You: What's bothering you?
Patient: Nothing.
You: Are you blind?
Patient: Huh?

You: Ok, what brings you here today? (You're pleased with that
    one.)
Patient: Sue and Miltie insisted I come.
You: Who are Sue and Miltie?
Patient: Those are the people who are living with me and pre-
tending to be my parents. They look just like my parents but I
know it's not them.

You have discovered what is called *Capgras Syndrome.* Further
questioning would show that the patient was not delusional. People
and things look to a person with Capgras Syndrome the same as they
did before, yet the patient insists they are imposters. If you point out
how strange it is that something or someone could look exactly like
your house or your mother but not be your house or your mother, he
or she will be the first to agree—but nonetheless insist they are
imposters. It is as if the visual information of an object has become
detached from the meaning or emotional content that usually accompanies it.

Table 1-1 summarizes the fourteen disorders that appear on the
slips of papers. See also Figure 1-1 facing the front of the chapter.
Think about how you might have behaved if you were acting out each
one. Think also about what kind of clever diagnostics you might try
that would have uncovered the deficit, especially when you do not
know what to look for.

Which brings us to the issue of trying to identify a deficit without
a list of possibilities. People feel lost without a list. The problem with
having a list is that if you think you know what the disorders are then
you may only find what you are searching for. This, in turn, may
cause you to miss the true nature of a deficit. For instance, suppose a
patient had prosopagnosia for everyone except Kevin Federline. If
you knew about prosopagnosia in advance from Table 1-1, and you
saw your patient had trouble with several faces, you might jump to
the conclusion simply that he had prosopagnosia. You wouldn't even
think to try Kevin Federline, because you did not know about a

Table 1-1 Disorders Affecting Perception

Disorder	Deficit in...	Site of Brain Lesion
Blindsight	Knowing one can see; perceptual experience	Area 17
Object agnosia	Naming, using, or recognizing objects	Areas 18, 19, 20, 21 on left side of brain and corpus callosum
Prosopagnosia	Recognition of faces	Areas 18, 19, 20, 21, 37 bilaterally
Color agnosia	Association of colors with objects	Areas 18, 19, on right
Agnosia for sounds	Identifying meaning of nonverbal sounds	Areas 42, 22, on right
Asterognosis	Recognition of objects by touch	Areas 5, 7
Autotopagnosia	Localization and naming of body parts	Area 7, possibly 40, on left
Synesthesia	Seeing sounds and other modality confusions	No brain damage
Alice in Wonderland Syndrome (micropsia)	Objects appear the wrong size, often smaller; shape and other distortions	Associated with migraine sufferers; post viral
Hemineglect	A region of space, usually the left, is completely ignored	Parietal lobe, on right
Capgras Syndrome	People and things looked to be the same, but believed to be "imposters"	Possible disconnect between temporal lobe and limbic system
"The man who fell out of bed"*	See Sachs, 1998* (patient thinks leg is not his own)	Temporo-parietal junction, on right
Alexia	Reading and recognition of printed letters	Various—occipital, temporal, inferior frontal lobes, on left
Anomia	Unable to remember names of things	Various—parietal lobe, temporal lobe

disorder "prosopagnosia minus Kevin Federline". The theories guide what you look for. By not having a cheat sheet ahead of time, you may be able to keep an open mind and be more flexible in your tests before reaching a conclusion. You may even be at an advantage over well-schooled professionals. If a patient seems to have agnosia for objects, it is possible she might still recognize real elephants. But you may never find that out, unless you bring an elephant into the room. You only find what you search for and having a catalog of disorders can get in the way of that. Knowing a universe of pre-existing disorders saves time, but comes at a cost.

Perceptual Learning

So, what is the connection with perceptual learning, the topic we are interested in? First, the disorders show that perception can be changed. Before the stroke, or the fall, or the gunshot wound, perception was different than it is now. This shows that there is nothing given or fixed or absolute about what you see or hear, if you see or hear at all, what you attend to, what you recognize, what you feel about what you see, and whether you even feel your limbs to be your own. However, these changes are occurring when perception is abnormal. When perception remains normal, but nonetheless changes, is the subject of subsequent chapters.

Second, we can ask: Do these disorders even involve learning? What is your intuition? Pause for a moment to access your gut sense. If you said, "yes" because you think that hopefully treatment will improve these poor souls' ability to function, then I need to clarify. Just consider getting and having the disorder, not what might happen afterward. Most people then have an intuition of "no", getting one of these or any disorder is not learning. But why not? The disorder affects your perception and you were not born with it. If you weren't born with it, why not then view it as learning? The answer is that the disorders make you *worse.* "Learning" refers to mechanisms that evolved to make you better. To paraphrase psychologists Paul Rozin and Jonathon Schull, who in turned paraphrased ethologist Konrad Lorenz, isn't it funny that learning always makes you better?[1] This assumption is so obvious a part of many people's intuition about learning that it often goes unstated.

But when this assumption is made explicit, it can lead to an objection, although I have found this objection surprising. If you discover Dad is having an affair, you have just learned something but does it make you better? How about if you find out you are adopted? Or you learn math but hate it. How do these things make you better? "Better" does not mean *feel* better but rather work better. Not knowing something that is true is an error and the capacity to learn the truth can then fix the error. Learning is adaptive, even if your emotions do not always appreciate the truth. If we add perception, we can have a working definition of perceptual learning. *Perceptual learning refers to any change in perception brought about by experience that fixes errors or otherwise improves perception*[2]. This is a very broad definition covering many things that are acquired perceptually in a lifetime. This will allow us to cast a broad net and therefore not miss any potentially relevant perceptual change. Its breadth is in contrast to previous interpretations of perceptual learning in the literature which have arbitrarily narrowed the field to one small subset of change. A broad approach should allow us to achieve a comprehensive unified understanding of how perception is changed by experience.

As an aside, how would you evaluate synesthesia? It does not appear to result from any brain damage that we know of. At least parts of these unusual percepts may not be present at birth, either. For instance, a woman developed a sour taste in her mouth whenever notes C and B were played together; other tastes were acquired for other musical combinations[3]. Is this learning? People's intuitions tend to come down to whether or not they believe that getting your modalities crossed, such as tasting musical notes or seeing colors when hearing words, will lead to improved perception. If yes, then learning feels involved. Those who believe instead that the synesthete is at a confusing disadvantage compared to the majority of perceivers tend to not think of synesthesia as learning. This is in keeping with our assumption that improvement aligns with learning. Incidentally, would you want to have synesthesia if you could choose to? Thirty percent of our respondents (90 out of 299 subjects) say "yes"!

So, in general, perceptual disorders resulting from brain damage are not perceptual learning according to our definition since they reflect something breaking rather than getting fixed or improving.

But they are related. We can say perceptual disorders and perceptual learning are *cousins on the family tree* (Figure 1-2). Disorders are still an effect of experience, and learning, too, is an effect of experience.

Experience

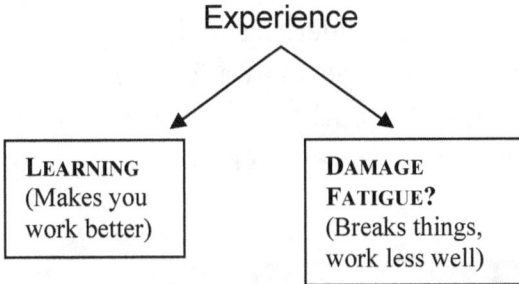

LEARNING (Makes you work better)	DAMAGE FATIGUE? (Breaks things, work less well)

Figure 1-2 The box diagram reflects the start of categorizing the effects of experience. Learning and damage are first cousins on the experience family tree. Is it an oversimplification to describe "experience" as one thing? If so, learning and damage may be more like second cousins.***

Perceptual disorders may still have some features in common with perceptual learning because there are only so many ways that things can change. There have certainly been some neuropsychologists who believe that disorders show how things come apart, not just fall apart. They hope the study of brain damage can carve the normal mind into its component pieces and thus teach us something about how the mind is organized in its undamaged state. These investigators subscribe to the dropped watch model: If you drop a watch a distance from the floor, it breaks into component pieces rather than random shrapnel. But we do not always know if things in the brain fall apart in rule-governed ways. Trying to infer mental components or organs from damage is also subject to logical problems involving what are known as dissociations and double dissociations[4]. While evaluation of these issues is beyond the scope of understanding perceptual learning, from our perspective, there is a more modest, achievable, and logically sound use for studying disorders. Disorders do not necessarily have to reveal mental organs. However, we can ask if any of the changes that occur in disorders match changes that occur when perception is working normally. We can use perceptual disorders to inform perceptual learning.

Anyone interested in perceptual learning should be on the lookout for unusual perceptual disorders. When one occurs, use it to guide research on people with normal brains and ask whether we should be looking for an analogous milder change in people without deficits. Anyone who thinks they have an idea for connecting a perceptual disorder to a normal perceptual change, please let me know. To learn more about how normal perception is changed by experience, read on.

_____

Why Growth of Your Body Requires Perception to Change: The Case of Sound

C lose your eyes. WAIT. I forgot that you cannot hear me with your eyes closed. Read these directions first. After closing your eyes, identify something you can hear—right now I can hear humming from the refrigerator—and then point to it with your eyes still closed. Open your eyes and see if your arm is aiming in the right direction. Try it again with something in a different place. If you hear birds outside, point to one, open your eyes, see if you are accurate. You probably are pretty good at finding the location of the object that is noisy with just your ears. This is true whether the sound is annoying or enjoyable. It is not a completely fair test because you may just be remembering where the refrigerator is. If you have someone with you, a better test of sound localization is to have that person quietly walk somewhere and make some noise, then you point to their location. Repeat twice more from different locations.

I hope you will really try the experiment even if you think it is obvious how it will turn out. Experiments often have results that differ from expectations. They also frequently reveal details that were not anticipated. Perhaps you will find you are more accurate with sounds on your right side than your left side. You can't know until you try. The best reason for verifying ideas yourself may be that it produces the frame of mind to always challenge what you hear. Otherwise, by assuming that what someone else says is right for simple claims, a bad habit develops that prevents independent thought when it really matters. I hope you will forgive the preaching, but I think it is important: Always evaluate things for yourself at each step in the process to assess the extent to which you reach the same conclusion.

Most people are reasonably good at determining where a sound is coming from with their eyes closed. "Marco Polo" makes a game of it. One person shouts Marco with his eyes closed in a swimming pool. Other players respond "Polo" and are hunted down through sound localization skill alone. Of course, before food came out of a BPA-lined tin can, your dinner made noise. Hearing faint footfalls or a rustling bush allows you to find the rabbit even if she's visually camouflaged. Lions that roar from a distance too far to be seen are best avoided, but you have to hear which direction to avoid. Substitute a banshee-screaming ambulance for a roaring lion to convert to modern society. Localizing sound is a good ability to have.

I have to confess that not that many people get stirred up about finding the sounds they hear in the environment. Why start a perceptual journey with such a dry topic? Maybe put it at the end or hide it in the middle, anything but at the beginning. Finding sounds will provide a clear case of changes to perception in the absence of any deficit. If you can master this, then you know you will be able to apply your logic to cases of perceptual change that are fuzzier. There's plenty of hiding that can be done behind fuzzy topics, just like rabbits behind bushes, but no hiding is possible for sound localization and as we'll see, growth. There are graphs, facts, technical details, and an equation or two. I also cannot get rid of a bad habit I picked up when a student at the University of Pennsylvania: Be sure to drive away most of your audience with something really hard and boring at the beginning so that only the "worthy" few remain. Remember, if it is just not for you, there is something for everyone in the area of perceptual learning.

Before demonstrating that perception must change, consider first just the way in which sounds are localized. These facts have been known seemingly forever. I will show how straight forward—and clever—it is to localize sounds. Jump to the next section if you already know all about it.

Backstory

It is important to have two ears for sound localization. Consider why: Suppose something is making noise on your right. A sound traveling

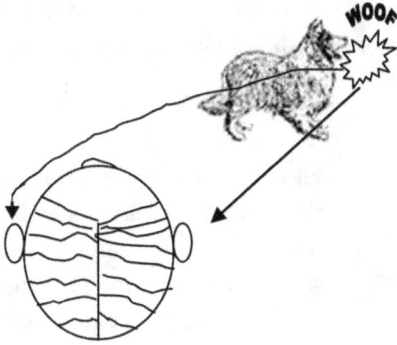

Figure 2-1 The sound from a barking dog will reach the closer ear slightly before the further ear because it does not have as far a distance to travel.

from the right will have to travel different distances to the two ears. Note how it is a shorter distance to the ear on the right because it is closer to Lassie, the source of the sound (Figure 2-1). So her bark arrives at one ear, the closer ear, before it arrives at the other.

Now consider: If Lassie is instead directly ahead, her alert (to tell you that Timmy is trapped in the well again[1]) will arrive at both ears at the same time because it is equally distant from both ears (Figure 2-2).

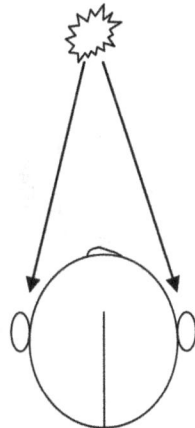

Figure 2-2 A sound straight ahead of your nose will reach the left ear and the right ear at the same time. The time difference between the two ears will be 0.

So depending on where the sound source is located, the amount of the time difference between the two ears will be different. That is the essence of how we know where the sound is coming from. There should be a satisfying sense of "Ah, I get it". If not, rinse and repeat.

To elaborate, when the sound is straight ahead, there is no time difference between the two ears. If the sound is slightly on the right side, the right ear will get it slightly faster than the left ear and there will be a small difference in the time of arrival of the sound between the two ears. Move the sound further to the right, and the difference in the time of arrival between the two ears will increase. That time difference will keep increasing as the sound keeps moving further to the right. It will be at its largest when the sound is maximally far from the left ear, or in other words, when the difference in the distance from the sound to each ear is at its largest. This will happen when the sound is directly off to the side, at 90°. It's easier to see with a picture (Figure 2-3):

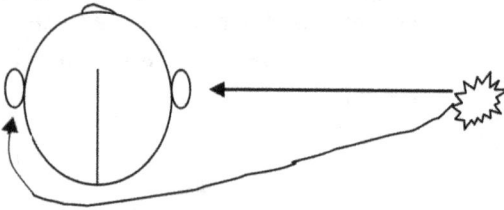

Figure 2-3 The direction of the sound is maximally far from the left ear. If the sound moves any further clockwise, it will begin to get closer to the left ear again.

Consequently, the time difference between the two ears serves as a cue as to where the sound source is located. Shown below is a graph of the exact time difference between the two ears for an average-sized head as a function of where the sound is coming from (Figure 2-4).

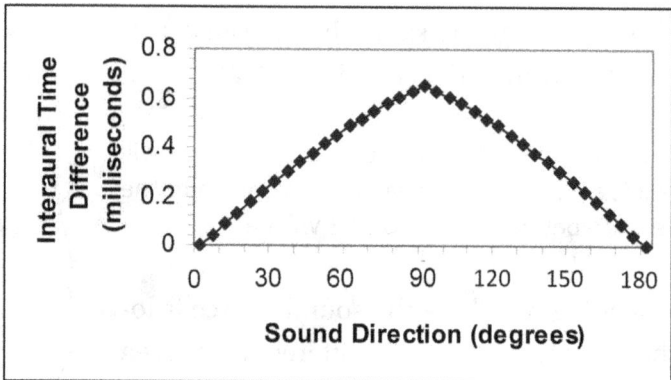

Figure 2-4 Difference in the time of arrival of a sound between the closer and further ear (Interaural Time Difference or ITD) for sounds ranging from straight ahead (0°) of the observer to his right side (90 °) to directly behind (180°). Numbers from Woodworth & Schlosberg[2] for sounds more than a few feet from the head.

The X-axis shows the sound's location in the horizontal direction.

Straight ahead is indicated by 0° while 90° refers to the location directly to the side (that is, perpendicular to straight ahead). Numbers in between are locations at smaller angles, e.g., 45° is halfway between straight ahead and to the side. Angles larger than 90° will be behind you; 180° would be directly behind you. The Y-axis shows the difference in the time of arrival between the two ears or equivalently, how much longer it takes the sound to travel to the further ear compared to the closer ear. As the angle of the sound increases from 0° to 90°, the timing difference between the ears increases.

I cannot help but point out that the timing differences are shown in fractions of a *millisecond*. If you are not familiar with milliseconds, 1 second is equal to 1000 milliseconds. Think about how little you can do in even 1 second. Even the fastest perceptual and cognitive abilities take a few hundred milliseconds. The auditory timing differences involved in sound localization are all less than a single millisecond! Much faster than the blink of an eye. Blinking actually takes about 200 milliseconds. When Woodworth, a pioneer of experimental psychology, first deduced these timing differences between the two ears[2], no one knew if such short amounts of time could be made use of. Now it is well known the auditory system can process these minute differences. Hopefully, you will find it intriguing that anything can be done in less than one thousandth of a second! Another trick of the trade is to find the intriguing things about every problem. There is excitement lurking in every topic.

Just a few more steps and we're there. Does the graph (Figure 2-4) suggest any difficulty in localizing sound? If it is not immediately apparent, consider a sound directly in front of you and a sound directly behind you. What are the time differences between the ears for each sound? They are *both* 0. Or suppose the timing difference between your two ears was .176 msec, how would you know if that meant the sound was coming from a 20° angle in front of you or from 20° behind you (160°)? Similarly, an interaural timing difference (ITD) of .3 msec could indicate the sound was either at approximately 35° in front or 155° in back. An ITD of double that, .6 msec, would indicate the sound was at 80° or at 100°. And so on for all the ITDs. You may have noticed these front-back confusions while driving. That screaming ambulance may be either ahead or behind you—it can be frustratingly difficult to tell. Sometimes this ambiguity is more generally known as the *cone of confusion*, a surprisingly good label that refers to every possible spatial position a given timing difference could be coming from. The cone of confusion for an ITD of .176 msec, for example, would include sounds at 20° in front, 20° behind, and all different heights at those angles. Note "left" and "right" would not be part of the cone of confusion. If the timing difference were .176, your auditory system would know if the sound was at 20° to the right because it would hit the right ear first whereas if it were located at 20° to the left, it would hit the left ear first.

Consider now one other effect of having two ears (besides wearing two earrings). A pair of ears provides a second way to localize the direction of sound. If you can't hear what a lecturer is saying from the back of the room, you would instinctively move your seat up to be closer to the sound. If a concert is too loud, move away from the band. Not only does the closer ear get the sound sooner, it also gets the sound LOUDER. The intensity difference between the two ears can therefore also can serve as a cue as to the direction of sound. Will it solve the cone of confusion? Nope. Interaural intensity differences are subject to the same ambiguity as the interaural timing differences. To determine if the sound is coming from in front or in back, one solution is simply to move your head. This repositions your ears for new timing and intensity information. It may not be wise to move your head too much while driving though, which contributes to a lingering confusion over the direction of an ambulance.

Timing differences and intensity differences both provide information from which to infer the exact direction of sound. These two systems excel at different frequencies of sound, that is, pitch, which ensures you can determine where a sound is coming from whether it is a low pitch purring or a high pitch scream. Yet there are many frequencies where both timing and intensity work, giving us redundant information about where the sound is coming from—a backup system. Redundancy is very common in perception and very important in perceptual learning. Whether the redundancy is already there for other reasons and is exploited by perceptual learning or is there precisely to facilitate such learning is a deep question that does not usually get addressed[3]. But I am getting far ahead and too abstract so I will return to the issue of how sound localization shows that perception must be changeable.

Sound Localization and Growing Children

Why should growth require that our localization of sounds be "plastic", that is, changeable? And suppose an infant were to grow overnight to adult size! Further suppose his sound localization system has not caught up to this new adult body. If a sound comes from his right side at a 45° angle, where will it sound to the giant baby to be coming from? And why?

You are actually armed with just about everything you need to

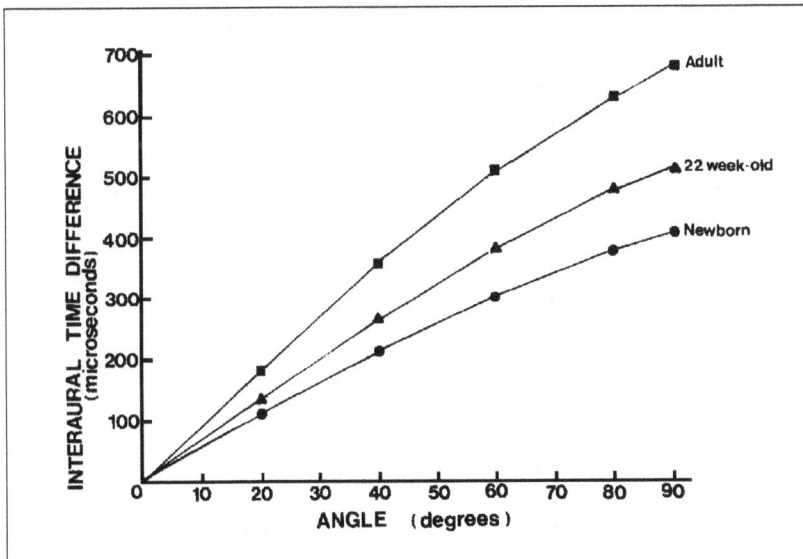

Figure 2-5 Time difference between the ears for infants and adults. 0°= sound straight ahead. Figure 3 from Clifton et al. (1988). Growth in head size during infancy: Implications for sound localization. *Developmental Psychology, 24*, 477-483. Published by American Psychological Association (APA); reprinted with permission.

answer these questions. Here's a hint. Consider what specifically is changing about the body because of growth that would affect sound localization. Then use those timing differences. Note I am asking you to figure out for yourself, the only way to true understanding, why growth affects sound localization. Students in my class are able to come up with the answers within half an hour, although they get to work in teams. Let me provide the rest of the information you would need to solve the details in the event you can find someone to indulge you for half an hour of discussion. Table 2-1 shows the same numbers as Figure 2-4 to make looking up ITDs for adult-sized heads easier and Figure 2-5 shows the ITDs for infants as well as adults.

Table 2-1 Time Difference Between the Two Ears

Sound Direction (degrees)	Interaural Time Difference (msec)
0	0
1	0.009
2	0.018
3	0.027
4	0.036
5	0.044
10	0.088
15	0.132
20	0.176
25	0.218
30	0.26
35	0.301
40	0.341
45	0.379
50	0.416
55	0.452
60	0.486
65	0.518
70	0.549
75	0.578
80	0.605
85	0.63
90	0.653

From Woodworth and Schlosberg, 1954, for sound more than a few feet from the head

When you return from brainstorming, remind yourself first of the sound reaching each ear when a sound comes from the right side (Figure 2-1). Then, if we compare the adult to the infant and line up the two right ears, we get the following picture:

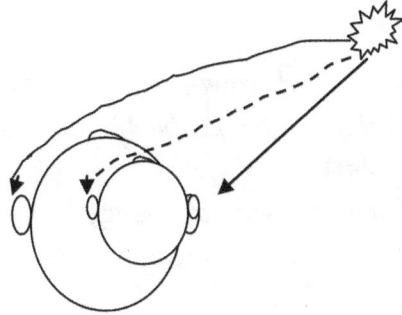

Figure 2-6 Arrival of the sound at the further ear for a small head vs. a large head. The sound has to travel a longer distance for the large head. This produces a bigger time of arrival difference between the two ears for the larger head for the same sound source.

If the sound reaches the right ears of the infant and the adult at the same time, it will take longer to reach the left ear of the adult than the infant because the adult's head is wider. Adults have fat heads literally as well as figuratively. Look at Figure 2-5 and find the interaural timing difference (ITD) for a 22-week-old infant when the sound is coming from 45°. It's about .29 msec. (Note this ITD is smaller than what it would be for an adult; a sound coming from the exact same angle would produce an ITD of .379 for the larger adult. You can refer to the table to get this number[4].) Normally, if the ITD were .29, the infant would localize the sound at 45° and everything would be fine. However, if the infant grows overnight to adult size and does not know it, then the actual ITD is .379 msec, because his head is now big like an adult. Will he localize it at 45°? No. In his "look-up table", 45° corresponds to .29, not .379. The ITD of .379 is bigger, so in his table, .379 must correspond to an angle larger than 45°—recall the larger the angle, the larger the ITD. The infant's auditory system is using a mathematical function appropriate for a smaller-sized head. To estimate what bigger angle the noise will sound like it is coming from, find the ITD of .379 msec (379 microseconds) on the Y-axis of the graph (Figure 2-5) and then scan across to the right until you hit the curve of the 22-week-old infant. Now look down to the X-axis to see what angle that corresponds to. The value is approximately 60°. That is, to a 22-week old infant, an ITD of .379 corresponds to a sound that is located around 60°.

Here is one version of a "full-credit answer" to the questions. The logic is more important than the actual number.

Why should growth require that our localization of sounds be "plastic", that is, changeable? Physical growth requires the localization of sounds be flexible because physical growth causes the distance between the ears to change and we know the distance between the ears is critical for localizing sounds.

Suppose an infant were to grow overnight to adult size! Further suppose his sound localization system has not caught up to this new adult body. If a sound comes from his right side at a 45° angle, where will it sound to the giant baby to be coming from? And why? It will sound to him to be coming from a bigger angle (of approximately 60°) to the right. The infant's head is larger but he doesn't know it. His localization is calibrated for a smaller size head. The sound at 45° actually causes an ITD of .379 according to the table, but in the child's scheme, .379 corresponds to a sound with a larger angle. (In the child's scheme, the head is smaller so the distance between the ears is smaller and the ITDs are usually smaller.) According to the graph, in the child's scheme, it looks like .379 corresponds to a sound at 60°

Physical growth causes the distance between the ears to change which, in turn, causes the same sounds to produce different timing differences between the two ears. Consequently, to accommodate growth there needs to be a change in the relation between the specific timing differences and the locations they indicate. Otherwise, using the old associations accurate for an infant's size would lead to errors for an adult size. Physical growth without internal change would cause you to search for something in the wrong location or to turn your head by the wrong amount as you tried to look at what was causing all the noise. You would not catch the rabbit.

Nothing is Ever Simple

Change in head size requires recalibration of the auditory system in order to maintain accurate localization of sound—but this is assuming infants can localize sound in the first place. Can they? If a sound was coming from straight ahead and another sound was coming from directly to the side, an adult would be able to tell they were coming from different directions. If they were closer together, with one sound say at 0° and one at 45° you would still be able to tell them apart. You can keep making the difference smaller and you would be able to distinguish them down to quite a small interval— only 1°. So move the sound by just a single degree to the left or to the right and an adult can detect it's coming from a different place. A young infant, however, can only detect the difference when the separations are a lot larger. There is steady improvement and by bumbling toddlerhood, he is considerably better but still not as good as an adult. This resolution ability in numbers is about 22° at 4 months of age, between about 12° and 19° at 6 months, 9° at 11 or 12 months, and 4° for an 18-month-old toddler[5]. I've saved newborns for last. They may only be able to distinguish left from right. If so, it is as if newborns have only 2 categories of space in audition, left side and right. A 6-month-old can divide space into about 12 pieces or categories (assuming a 15° resolution), a 12-month-old into about 20 categories (9°), an 18-month-old has about 45 categories, and finally, an adult arguably has 180 categories of space. Auditory localization is, therefore, also getting more precise or refined with development such that more and more slices of perceived space are possible.

Consider now our infant who overestimated the location of a sound because his head grew and his auditory localization system had not yet caught up. In our example, the target at 45° is incorrectly heard at about 60° for an error of about 15°. However, as just discussed, a 6-month-old makes errors of 12 to 19° anyway! Note that unlike the error caused by head growth, this latter error is not systematic. Head growth produces an error that is always of the same direction and size for a particular location each time a sound is heard. With this new problem though, the errors are instead unpredictable with overestimations sometimes and underestimations other times. There is more slop, less precision. But the consequence of this lack of precision is that the systematic error caused by head growth may not matter. His auditory localization may not be sufficiently fine-grained at that age to distinguish between positions that are 15° apart. If he can't tell that two sounds 15° apart are in different places, what difference does it make if head growth produces errors of 15°? To extend the reasoning to other ages, by 18 months, the child has a resolution ability of 4°. Thus, the amount of head growth would have to produce overestimation errors of more than 4° to be of real consequence. If not, it would be within the child's resolution ability anyway.

Does this mean that the change in head size with growth does not require that audition recalibrate itself after all? No. Let's go back to the young infant. When the sound is at 45°, the error of 15° following uncompensated head growth is arguably too small to be noticed. But the greater the angle, the greater the error will be. If the sound were at 60° instead, the error would be about 30° for the same head growth. This can be verified using the graph in Figure 2-5. A 30° error does exceed the resolution ability of the infant. At least for some sound directions then, if the auditory system were not recalibrated for physical growth, sound localization would show errors above and beyond the inherent noise of the auditory localization system at young ages.

So there are two things going in development over the same time period and both must be kept in mind: 1) The head is getting larger, which requires a change in association between each sound location and its interaural time difference. 2) There is an increase in precision, such that the resolution ability of auditory localization is becoming finer grained with development. The latter is not required by head growth, but occurs generally in development with maturation. Perhaps it helps to think of precision as an increase in the number of associations, whereas physical growth requires change in the content of the associations themselves. *It turns out these two processes we have now already seen in development, recalibration and resolution, mirror the two major classes of perceptual learning in adults as well, perceptual adaptation and perceptual expertise.* We will come back to this in later chapters.

An Article by Clifton, Gwiazda, Bauer, Clarkson, and Held

This sounds like the beginning of a bad joke. How many psychologists does it take to... The graph you have consulted for the calculations of infant timing differences (Figure 2-5) comes from an article by developmental psychologist Rachel Clifton and several colleagues. The paper is entitled "Growth in head size during infancy: Implications for sound localization"[6]. Credit for understanding the effects of head growth on changing sound localization belongs to the last author, Richard Held, a great and classic figure in perception and development who has been educating all of us for more than half a century. Back in the in 1950s, Held explained why growth requires recalibration; decades later, he and colleagues calculated precisely the differences in the interaural timing occurring at different ages that you see in the graph.

They also report in the article an experiment in which they measured the head size of infants—both the distance between the two ears and the circumference of the head—every other week from 8 to 22 weeks. They additionally measured separate groups of newborns, 3 to 5-year-old children, and adults. They found that head size changes rapidly, head shape changes, the rate of change is the same for boys and girls, and there is no difference from measurements that were taken on populations 60 years prior. The interesting aspect of their work, however, was not the experiment, but their conceptual analysis. Head growth is used in clinical medicine as a measure of appropriate development in infants. When parents bring their infant to a doctor for a check-up, the doctor may measure her head and compare it to published norms of her age group to screen for any problems. Consequently, there were plenty of head measurements that already existed that could have been used by Clfton and colleagues in the study. But their work may not have gotten published without the collection of new data; sadly, experiments are simply expected in psychology. The main finding was conceptual. As we have discussed, they showed that there needs to be an update of the map of auditory space because the old one would lead to errors.

To calculate the timing differences required for the different head sizes that you have seen in the graph, they used an old formula derived by Woodworth[2], too far in the past for me to remember his first name. They then showed how the timing differences are different for different head sizes, thus arguing that a different mapping is required between the sound positions and the ITDs at each different age. Woodworth's original formula was:

$$\text{ITD} = \frac{r}{34,400} (\theta + \sin \theta) \tag{1.1}$$

where θ is the angle of sound, r is the radius of the head in centimeters, and ITD is the difference in arrival of the sound at the two ears. The denominator is the speed that sound travels in centimeters per second. Woodworth derived his formula based on physics and geometry.

If you ever read the Clifton article, you will note they argue that without recalibration, the infant or child would *under*estimate the location of the sound. That is, it would sound to be coming closer to the child's midline than it really was. Recall, we concluded the sound would be overestimated, such that it would sound like it was coming from a more peripheral position than it really was. The reason the conclusions seem opposite to one another is that we made opposite assumptions about how people would be built if plasticity were not possible. We assume that the kid begins with an appropriate kid-sized map, and then ask what error would get made if he grows but does not update the map. On the other hand, the authors of the study made the assumption that the kid starts out with a map that comes pre-sized to fit an adult head, and ask what kind of errors the kid would make as a result. Another aspect to note is that while they calculate the ITDs at different ages as a function of the direction of sound, they do not calculate the actual errors in sound localization that would be made as a function of all the different directions of sound. This is something I needed for my own work and I went back to the original Woodworth formula to derive this new formula. I present it in the appendix here if anyone is interested. (See also Chapter 9.) To my knowledge, no one has calculated how these errors compare to the precision of sound localization at each age (the other thing that changes with age, as discussed earlier) and each sound direction. I ask my students to do this as a difficult extra credit problem because I like to query fresh minds about issues that we do not yet have answers to.

Perceptual Learning Framework

Let's go back to our definition of perceptual learning: any change in perception that results from experience that corrects errors or otherwise improves perception. We can look at how the developmental situation of head growth and sound localization relates to perceptual learning. In our familiar example, a sound is located at 45° but the rapidly growing superbaby with the adult head size mislocalizes it at 60°. If the auditory system finally catches up, he will now have the appropriate adult-sized associations between the ITD and the sound directions and will correctly localize the sound at 45°. Is this a change in perception? Yes. Initially, the object sounded like it was coming from 60°; now, following recalibration, it sounds like it's coming from 45°. Would this be an improvement? Yes. Hearing the sound at 60° was an error, 45° is correct. Does the change result from experience? Presumably yes. We will turn to this in the next section.

First, a framework that can help analyze any situation and assess whether it is a candidate for perceptual learning is very simply:

t1

Experience

t2

It's simple, and I often apologize for its simplicity, but it's useful. Compare the perception at one time (t1) to a later time (t2). If it's different given the same stimulus, and it results from experience, then it is a candidate for perceptual learning. I adapted or borrowed (or stole) this framework from Bob Rescorla, the undisputed master of animal learning, who uses it for Pavlovian and instrumental learning in animals. Although I always site him, I felt like a thief until recently when I discovered that he adapted or borrowed (or stole) a set of ideas for thinking about learning from Eleanor Gibson (see Chapter 6) written in the 1960's. Ironically, while Bob did not use them for perceptual learning, Eleanor Gibson did. Perhaps we are bringing these ideas back home where they belong.

In our example, the stimulus at both t1 and t2 is a sound at 45°, yet at t1 the perception is a sound at 60°, while at t2, the perception is a sound at 45°. Perception has changed.

How Do You Get Recalibration?

I have thus far been concerned with the errors that would occur in development if recalibration did not occur. Because neither adults nor children make large perceptual errors, we know recalibration does indeed occur. Before any recalibration, the object is at 45° to your right and it is perceived at 60°, to return to the example. But now imagine that you are the baby: You hear the dog barking and as far as you're concerned, it's at 60°. How would you know that you are wrong?

A researcher in vision, Marty Banks, suggests three ways that experience from the environment can be useful for recalibration in development[7]. He suggests that information can come 1) from "another system entirely" 2) from "different but closely related systems" or 3) from "internal discrepancies". These are all comparisons. That is, what if you, the baby, got to compare what you heard to something else and found the information didn't agree? This kind of internal war is so important that I devote an entire chapter to it later (see Chapter 7). Banks was working on a different development problem in vision, but we can also apply the three ways to sound localization. For sound localization, "another system entirely" would correspond to getting to compare audition to another modality, like vision. Suppose you also see the dog when it's barking—and it's at 45°, not 60°. That may give you a clue that something is wrong and that recalibration is needed. For a "different but closely related" system, that would correspond to information within the same modality as audition, but a different system. We have seen briefly that another system for sound localization besides timing difference between the two ears is the intensity difference between the two ears. Head growth need not affect these two systems identically and following head growth, the intensity differences may lead to localizing the sound somewhere other than the location specified by the timing differences. That disagreement would also provide a clue that something was wrong. Finally, "internal discrepancies" corresponds to information that provides contradictory information all within the same system. What kind of contradiction might that be for timing differences between the two ears?

One candidate is to consider what happens when the object is located at larger angles further away from straight ahead. Recall that that for a sound at 60°, the error for a 22-week-old infant who grew to an adult head size overnight, but did not know it, would be a large 30° error; he would hear the sound to be coming from 90°. Now think of the very interesting case of what would happen if a sound comes from a direction that is even greater than 60°? If you look on the adult-size table—our superbaby's head is adult size—you will see it produces a very large ITD (see Table 2-1). If you now compare these to the kid-size data shown in the graph (Figure 2-5), you'll find that the values are larger than anything the kid has ever experienced! If he looks to see what an ITD of .710 msec, for instance, is coming from on his look-up table that converts ITDs to sound positions, he will find that the value is not there. The largest ITD he has had is not much more than .5 msec, which corresponds to a 90° target. A larger ITD cannot correspond to a bigger angle, since we've seen the ITDs for bigger angles get smaller again. How baffling this must be for the infant! [8] Such an internal contradiction may clue in the auditory system that something isn't right and needs fixing.

Which of these three potential cues is the best trigger for change? Marty Banks favors the last, internal discrepancies, because he says the other systems that provide the contradictory information for the first two cues could themselves have errors. He suggests the standard against which one system is being judged may have its own growth problems. For our situation, what would happen if audition is compared to vision to determine if audition is correct but vision turns out to be wrong? This definitely is a problem, but the disagreements these comparisons between systems provide can nonetheless tell you something is wrong, even if it does not tell you exactly where the problem lies. In addition, perhaps the internal discrepancies that Banks favors are not sufficient cues for recalibration because there may be some situations in which there are perceptual errors yet no internal contradictions within a single system. So I take a broader view and say that, in principle, any of these can trigger the detection of an error, which in turn will lead to the necessary change in associations.

Which is really used in development? No one knows for sure. When adults need recalibration, it is the first approach, comparison between modalities, that appears to get used a great deal—at least in the laboratory. A lot more on this in subsequent chapters. Given the adult findings, the informative value of each cue, and the frequent use of backup systems and redundancy in perception and perceptual learning, the best guess is that all three are used.

If you were being a skeptic and I didn't distract you with all the details, you might raise the issue of whether much, or even any, experience is required for perceptual change. Consider something like "learning" to walk. Walking isn't really learning. The motor components of walking are considered rather to unfold in a preset order as the infant matures, regardless of the particular types of experiences encountered. Maybe recalibration is also maturational and occurs at a genetically determined preprogrammed sequence like walking. Why is this a bad model of recalibration for growth-induced changes in the distance between the ears? Because there is no way to know ahead of time how much your head grows. Nutrition plays a factor in growth, not just genes. The only way to have an accurately calibrated auditory localization system is to use experience to fine-tune the details to the exact fat head size that you happen to develop.

Two Final Considerations

Growth in the real world is slow. There are no superbabies that get huge heads overnight except maybe in the movie "Big" or the original television series "Dark Shadows" or the defunct soap opera "Passions" [9]. Does this change the need for recalibration? If you said "yes", you are in good company—but wrong. Many of my colleagues even are under a false impression that very gradual continuous growth means that recalibration is not required. Slow growth does not change the need for recalibration. One way to think about it is to consider how recalibration interacts with the precision issue. The errors involved in slow growth will initially be small. If the size of the error is below the auditory resolution, the error may not be detected. If the error is not detected, it will not get fixed and will accumulate. As growth continues, eventually the error will accumulate enough to exceed the auditory resolution. In general, if growth is too small to either need or get recalibration, eventually the cumulative error from continued growth will require that growth be accommodated or else there will be big systematic errors in perception. If instead recalibration can occur for very small errors that are within the auditory resolution ability, then all the recalibration issues discussed are also identical, just with smaller numbers than the examples provided.

Finally, it is important to note that growth during auditory localization is not an isolated problem. What else is growing besides the distance between the ears? The distance between the eyes gets a little bigger and the distance between the shoulders gets a lot bigger. I am singling these out because they are changes that clearly affect perception in ways that have been well worked out also. The former will change how you perceive depth—something that Banks developed— and the latter will change how you reach for things—something that I developed. Perceptual learning is needed during growth. Without it, perception would cease to give you the accurate information that you rely on. I hope you don't just memorize this conclusion. Convince yourself it is true, or come up with reasons you do not agree if you do not.

I firmly believe that the changing of perception with growth is a realization about development that every educated thinking person should take with them for a lifetime. Whenever you see a kid, you should now and forever know and think about how his perceptual system will have to change as he gets big.

If one were reducing all the discussions to a single take-home message, that message would be: *Perceptual learning is needed to maintain accurate perception in the face of physical growth.* Whether this is also generally true for adults who have stopped growing leads to the next chapter.

Harry Potter's Holly and Phoenix-feather Wand

Whatever position you are in now, close your eyes and do not move. Wait; I did it again. I'll have to tell you what to do first, although it works better if I can talk to you while your eyes are closed. And don't forget you have to actually do it. After you close your eyes, imagine where your left foot is. Feel your foot and where it is. Now start going up your left leg as if you were outlining your body. Go up the outside of your left leg, up the hip, along the trunk. Make sure you feel each part of your body before continuing. Keep going up. Feel along the outside of your arm to the shoulder, around the shoulder, to your neck, and around the top of the head. Go down the right side in the same way, ending with the big toe of your right foot. I'll throw in one sobriety test before you open your eyes, which I hope you have never had at the side of the road. Put your right arm out to your side then touch your nose with your index finger.

You should have been able to feel where each part of your body was. Your left arm might have been bent at the elbow and your left leg might have been crossed over your right leg. Whatever unique position you happened to be in, you were likely able to mentally trace much of the outline of your body. You can probably draw it on paper if you are artistically inclined. *Body schema* includes keeping track of the position of the body at all times. You have to keep track of where your body is both when you are still and when you move—even though you are not normally consciously aware of doing so. This ability is adaptive. It kept your ancestors from decapitation by tree branch, not to mention allows you to find your own nose with your index finger.

Body schema is already interesting, but becomes even more so when you consider tool use. Not only do we use tools, but they seem to become extensions of our bodies. In the Harry Potter novel series[1], we find that:

"He picked up the holly and phoenix wand and felt a sudden warmth in his fingers, as though wand and hand were rejoicing at their reunion."

And in the Hunger Games trilogy[2]:

"He'd been on boats his whole life. The trident was a natural, deadly extension of his arm."

Objects we manipulate repeatedly for a set purpose seem to become part of the body schema. If you are taking notes with a pen, it almost feels as if the finger has grown to include the tip of the pen. Try tapping with the pen on a table. If you direct your attention to what that feels like, doesn't it feel as if the tapping is occurring where the pen meets the table? If you do not find that peculiar, consider that you cannot have sensation in the pen—it is not actually alive or a part of you. You do have sensory endings in the tip of your finger though and that is where the tapping sensation really comes from. Yet we normally misplace it to extend to the pen. Another example of extensions to body schema comes from sports equipment that feels to become a living part of the body, especially for good players.

Think now of one thing that you use that seems to have become an extension of your body. I like to go around a room with one extension per person until there aren't any more suggestions. In addition to sports equipment like tennis racquets, hockey sticks, and softball bats, I have gotten reports of a cello that felt like an extension of a musician's chest, a sword for a martial arts practitioner, a 10-ft long grapefruit picker during a summer job, a hammer for a part-time carpenter, and a small shovel for a gardening enthusiast, though only when she wore gardening gloves (which she felt naked in the garden without). I have been getting claims with increasing frequency about cell phones, although it is not clear if cell phones become extensions like other objects. Still, one woman said she felt like a part of her was missing when she accidentally left her phone behind. Anything that can produce that feeling is worth considering as a potential body extension. One of my own anecdotes concerns a car, a common source for an extended felt-body shell. One day at an intersection, a lazy driver did not pull up to the intersection in preparation of making a left turn. In Arizona, left turns are made after the green light. During busy times of day, there is no time to make the left before the light turns red unless cars pull up ready to go. I honked my horn as a gentle reminder, but there was no sound. The horn must have been broken, but my immediate sensation was that I felt like I had actually lost my voice. I was distressed at not being able to "speak", just like I felt when I really did have laryngitis years earlier. The car had become an extension of me to such an extent that its voice became my own.

Another body-extension experience that I had was similar in some ways to pen tapping and involved an ice cream cone I was eating near a big decorative fountain. The fountain would occasionally spray droplets of water, but I was certain none of the dirty water was hitting my ice cream. The reason I was certain was completely irrational. I assumed that had it been hitting the ice cream I would have felt it because I could feel it when it hit my hand! Since I didn't feel the water on the ice cream, I figured it must not have reached it. That is, of course, an absurd belief since there are no touch receptors in ice cream. The cone became a part of me and I made assumptions about it based on how the rest of me works. I knew logically for both the car and the ice cream that my conclusions were not right, but that is what they felt like and the feelings were very strong.

Here is one very recent event, which you have the benefit of hearing about before anyone else. I bought a mechanical reacher for picking things off the floor because I hurt my back (Tip 4—we are up to number 4—Do not get old, to which an even older mentor of mine, Henry Gletiman, adds: "Do not die"). A reacher is like a grapefruit picker, only a lot shorter. You squeeze the handle at one end and a claw a couple of feet away at the other end closes. After I used it probably a hundred times over the course of a couple of weeks, I spontaneously did something different than usual. When I used the reacher to grab a napkin off of the kitchen floor, instead of putting the napkin in my free hand like I had been doing and then placing the napkin on the counter, I put it on the counter directly with the reacher. I did this even though the counter was in easy reach of my actual hand and did not require the reacher. With practice, somehow I began behaving with the reacher exactly how I would with my own hand. If I picked something off the floor pre-injury, I certainly would not transfer it to the other hand before putting it on the counter. It's no Harry Potter magical wand, but nonetheless, arm plus reacher became my new much longer arm and hand.

Examples of tools that become extensions of our bodies suggest that our body representations are plastic. If you think about it, they have to be plastic. One reason is because of growth. In the last chapter, we saw that growth of the head would produce errors in sound localization were the new supersized you not accounted for. Growth would also produce errors in body schema if not accounted for. If you tried to navigate around the world believing your body was smaller than it really was you would bump into people and get stuck in all sorts of narrow openings. The body schema also has to be plastic even when fully grown simply because we move around frequently. Just bending your elbow requires the body schema to change if it is to accurately capture where you are in space at all times. Otherwise, you would be wrong about where parts of your body were located and your protruding elbow could take out an eye. Finally, returning to tool use, a plastic body schema may allow effective tool use. Greater proficiency at using a tool goes along with a felt alteration in the body schema to include the tool. One of my undergraduate students used this idea to train BMX bicycle racers to improve their speed through exercises designed to make the bicycle feel like a seamless part of the body, like a modern-day centaur. Another student said she watched so much TV, she really did feel like the couch had become a part of her. I'm not so sure she meant that literally, but perhaps she's onto the future of the American body schema, half human, half couch (Figure 3-1).

Figure 3-1 Ultramodern centaur? We spend so much time connected to the couch, perhaps it will soon become part of the body schema. Original cartoon by Felice Bedford.

Have a look at a stick figure that undergoes growth, movement, and a tennis racquet (Figure 3-2). If you were to mentally trace the outline of each figure like you did for your own body earlier, note how the body envelope changes size and shape. It's an obvious point, but attaching a visual image to it makes it concrete.

Figure 3-2 From left to right: Body envelope changing from the original due to growth, movement, and an extension from a tool. The size and/or shape of the figure's outline changes.

Body Schema in the Brain

I usually do not spend much time on findings in the brain. There is an explosion of such research relevant to perception, but most of them have been slowing progress on how the mind works rather than facilitating it. But this finding is striking. I do have a question though as to whether you think this is a noteworthy finding, since I can probably still be convinced otherwise.

Let me start with the basic property that there are neurons in the brain that respond to more than one perceptual modality, such as to both vision and touch. Some of these neurons become activated whenever any place on the hand is touched. Such a "hand neuron" is specific to a hand and does not fire if any other part of the body is touched. The same neuron also responds if it detects something visually—not just to any visual stimulus, but only those in the same place as the hand. For instance, if you shine a flashlight onto the index finger, the neuron will fire. One fun finding is that if you move the hand, the visual locations that the neuron responds to moves with it. It really is the hand that excites the neuron and not just where the hand happens to be. Another instance of a bimodal neuron is one that responds to touch stimulation at the shoulder and neck region and responds to all visual stimuli in locations that would be in reach of the arms.

Angelo Maravita, a psychologist from Italy, and Atsushi Iriki, a cognitive neurobiologist from Japan, collaborated to ask what these bimodal neurons would do following tool use. They reported their results in an article entitled "Tools for the body (schema)" in the journal Trends in Cognitive Science[3]. They trained macaques first for two weeks to use a rake to pull in food they could not reach. This initial training was essential because the macaques, unlike humans, do not spontaneously use tools. Once trained, they recorded the responsiveness of bimodal neurons in sessions before using the rake and following 10 minutes of using it to retrieve food.

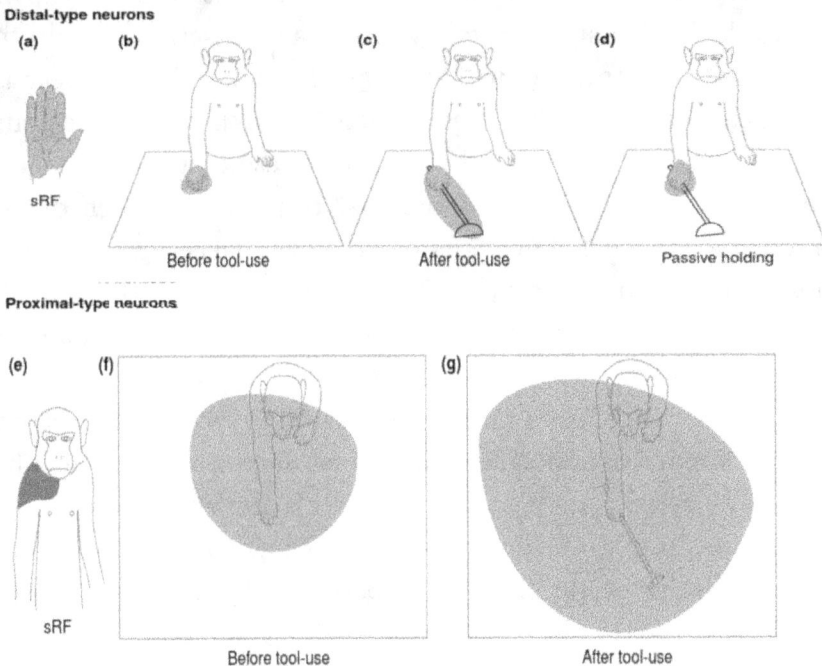

Figure 3-3 Changes in bimodal neurons from the pre-motor cortex before and after tool use. Shading indicates region that causes the neuron to fire when touched (Panels a, e) or visually stimulated. Top row: hand neuron Bottom row: shoulder neuron. sRF=somatosensory receptive field.
Figure 1 from Maravita, A., & Iriki, A. (2004). Tools for the body (schema). *Trends in Cognitive Sciences*, 8(2), 79–86. Published by Elsevier, LTD; reprinted with permission.

They found that the visual receptive field of some of the bimodal neurons, that is, the regions where visual stimuli excite the neuron, actually expanded to include the rake! (Figure 3-3) Previously, only a visual stimulus (like a flash of light) that occurred around the hand would excite the neuron, but after using the rake for a few minutes, the flashes of light around the rake did so as well. Just passively holding the rake in the hand but not using it to retrieve items did not lead to an expanded receptive field (see panel "d" in Figure 3-3). Also note that it is the visual receptive field that expands to include the rake, not the somatosensory receptive field. The latter cannot expand to include the rake since the rake cannot have sensation, just like the ice cream cone or pen cannot have sensation. I mention this because it is an easy error to make unless explicitly thinking about the point. The investigators also found the same basic result for another kind of bimodal neuron, namely the one that responds to touch on the shoulder and neck. Following practice, the visual receptive fields of those shoulder neurons expanded so that now everything in the extended reach of arm plus rake would excite the neuron and no longer just the region within arm's reach.

Is Maravita and Iriki's finding of expanding receptive fields with practice important or not? After all, everything we do or perceive has a neural correlate, so is there any reason this particular neural correlate is interesting? If you don't know what I mean by neural correlate, all I mean is that everything we do—perceive, think, feel, act—has to have neurons in the brain somewhere doing something that goes along with the activity and makes it possible. That does not necessarily mean it indicates anything about the psychological process. This particular finding, however, seems to suggest something at the psychological level of explanation above and beyond just being a neural correlate. It seems to be indicating something about body schema. How do we know that tools become an extension of the body? So far, what we have discussed is that it *feels* that way, that's why. With practice, tools feel like they become part of the body. The anecdotes and the feelings are fascinating and should not be dismissed, but they are difficult to quantify—that is, measure. It is also difficult to use them to distinguish between a tool that becomes incorporated into a body schema and simply getting good with a tool but without any changes to the body schema. The neural finding seems to provide evidence that tools really are extensions of the body schema.

This is because everything is happening within the same neuron, both the rake and the hand. Suppose we just looked at the activity of the neuron, without looking at what triggered it, and watched it go from a resting state to rapid firing. We would know it was stimulated, but would not be able to tell if it the cause was a visual stimulus around our own body or around the new extension. It is as if the hand and the rake have become indistinguishable, which provides evidence that the tool has indeed become incorporated into the body schema.

Suppose the neural correlate had turned out to be something else. For instance, suppose instead of expansion of receptive fields, nearby neurons were recruited such that practice with the rake caused these neurons next to the original one to fire whenever there was a visual stimulus around the rake. Would such a "nearby neuron" result speak to the issue of whether the rake became an extension of the body? This is the question that has to be answered to decide if the neural finding is interesting for psychology. This alternative correlate, had it occurred, would not really inform what happens to the body schema with tool use. If the hand and the rake were served by different neurons, there would be no evidence that the hand and the rake had become indistinguishable. Nor would it prove that the body schema had *not* changed to include the tool. Rather, a "nearby neuron" result would indicate that at the neural level with these particular neurons it is simply not apparent what is happening at the psychological level. This alternative finding, had it occurred, would still answer the question "what happens to neurons with extensive use of a rake?" but that is a different question than "what happens to the body schema with extensive use of a rake?" In general, whenever you hear about a

neural or brain event that co-occurs with perception or behavior—
and there are tons of them these days—ask yourself whether that
indicates what the neurons are doing (biology) or whether it tell us
about the actual mental processes (psychology).

Going back to changes in body schema from experience, are expanding neural receptive fields the only way to get objective evidence
that tools become extensions of the body? One behavioral method
makes use of an anomaly of body schema. When I told you to shut
your eyes and mentally trace the outline of your body, I mentioned
that if your legs happen to be crossed you would be able to accurately
portray that shape. It turns out though that when we cross any body
parts, whether they be legs, arms, or fingers, there is a certain amount
of sensory confusion that occurs. Try the following, although do not
get your hopes up since it is not an earth shattering-demonstration.
Cross your middle finger over your index finger, close your eyes, and
gently poke with something pointed, like the tip of a pencil. Decide
which finger feels like it has been poked; now open your eyes. Were
you right? At the very least, there is less certainty about what's going
on there than if you repeated the experiment with the fingers un-
crossed. This is a variant on a cross-finger demonstration reported by
Aristotle—such a long time to have known this. To turn this anomaly
into a methodology, imagine a probe that vibrates each of a subject's
hands and she has to judge quickly which zap came first. It turns out
the judgment is disrupted if the hands are crossed at the wrist.

I'm not sure if you can see where this is going. One can now ask:
What will happen if the *tools* are crossed? Suppose practice has
occurred with two rakes, one in each hand. Suppose further that you
are instructed to cross the rakes—but not your hands. If each rake is
now vibrated (which of course you will detect in your hands, not the
rakes, even if you're not aware of that—remember the pen or the ice-
cream cone), will the judgment be altered because the tools are
crossed or normal because the hands are not? It turns out it's the
former! Temporal judgments are disrupted when the rakes are
crossed just like if your hands are crossed. This finding suggests that

tools come to be processed just like your own hands, warts and all. So in general, we should be able to exploit any kind of peculiarity of normal body schema to see if tools acquire that same peculiarity. If it does then it would provide behavioral evidence that we incorporate tools in the schema above and beyond that it feels like it does and without having to put electrodes in the brain. Besides crossing body parts, adopting unusual postures also appears to produce sensations that are different than the regular body schema. In principle, this could be exploited in a test as well. Overall, however, very few behavioral measures exist for convincingly demonstrating, much less quantifying, changes in body schema.

Design an Experiment

First try the following task. Draw a horizontal line on a blackboard (improvise...) and get a longish stick, such as a tree branch, or your mother's reacher, or even a ruler. Stand away from the board, such that you could not reach it just by extending your arm, and use the stick to "draw" a vertical line that bisects the horizontal one exactly down the middle. Now you need a laser pointer, which hopefully you have around to torment your cat with (they chase the spot of light). If you can't get a laser pointer, try a flashlight or the reflection from a small mirror. Stand at the same distance from the wall that you did for the stick and now perform the same task of bisecting the horizontal line with the spot of light. Did the beam of light feel like an extension of your hand? What makes the laser different than the stick? Most people feel like the stick will be easier to become an extension of the hand for performing tasks like bisecting lines or writing. Must objects be physically connected to our bodies in order to become extensions?

What are other questions like this? How would you explore the limits of changing the body schema with extensions? Design an experiment that asks a new question about the plasticity of body schema that you would like to know the answer to, like the question on whether objects must be physically connected to our bodies in order to become incorporated into the schema. When I ask under-graduate students to come up with an experiment, I am often sur-prised that they do not always know how to do this. Too many people are used to being told things, but those things are just the tip of the iceberg about what is known and, even more importantly, could be known. This is often a turning point in someone's education that determines what kind of scholar, or indeed human being, he or she will be. I get a number of students that say they understand all the experiments and other material I have told them about, but really don't know how to go beyond that on their own—or even what it means to go beyond it. If this describes you, then, whether you are a student or not, you should work on it until it becomes second nature. Anything you encounter should reflexively make you think: "Ok, that's interesting (or not), *but now what happens when you do X*?" It will take a while until you can design all the details of an experiment that are free of pesky problems, but the basic instinct to question requires less experience and you should be able to generate at least a half dozen possible variant questions about the basic phenomenon.

Right now I only need one. Students usually get the better part of a class period to think of a variant and how they might test it. In addition, they get to work in a group like they did for sound localiza-tion and growth. Have a mocha latte and come back when you have stewed about an experiment.

Some people wanted to know if a vertical extension to the body would be better incorporated into the body schema than a horizontal one. They thought vertical extensions might be easier for women because of previous practice with high-heeled shoes. They suggested comparing swim flippers to stilts and using the crossed tool tech-nique as a way to assess if the extensions had come to behave like the actual body. Many groups proposed using the crossed tool technique to assess if the extensions become part of the body schema, since it can be difficult to come up with a convincing test that doesn't require brain surgery. Some students tried to develop such tests, including questionnaires that would ask participates to rate how strongly the object felt like a part of themselves, and behavioral measures like how successfully they touched their own nose. Another student designed an experiment using a glass rod and Crisco cooking oil. She reasoned that since the oil has the same refractive index as glass, it would make the rod invisible and she would be able to determine if a tool needed to be visible to become an extension of the body. Don't be concerned if this never would have occurred to you—ever—as this brilliant student went to graduate school to invent her own field on the mathematical modeling of brain functioning.

Some people took their questioning in the direction of practical applications that could be useful for real life problems. One group of students wanted to know if a prosthetic limb would be easier to incorporate into the body schema when it was on the dominant side of the body. Another group wanted to know if the belly of a pregnant woman would be accommodated as part of her body schema. They designed an obstacle course with different sized-openings to try to see how good the women would be at automatically navigating through the right-size opening compared to women who were not pregnant. There are issues of interpretation here, both if the women succeed and if they do not, but it was a nice creative start. It is indeed also quite similar to something a colleague of mine researched a number of years ago when his wife got pregnant and he got curious.

Here are some other variants it would be interesting to know: Must the tool be rigid like a stick to become an extension or can items like string work, too? Is there any limit to the length of an extension? It seems like shorter objects would be easier to become extensions than longer objects; is this true, and if so, is there some length at which the process of body schema expansion breaks down? How about a limit to the number of extensions: Could we have 2 new ones at the same time? How about 20 new ones? Would extensions that were symmetrical on both sides of our bodies—like two rakes— work better than just one since our bodies are vertically symmetrical? Does the shape of the extension matter? Perhaps if the shape were similar to something already on our body that would make it easier to incorporate. How about the weight; as long as you can lift it, can it become an extension even if it's heavy? For any of these you can ask if it poses an absolute constraint, such that it can never become an extension, or is just harder and would take longer. Which leads to another kind of questioning: How long does it take for an object to become an extension? Are we talking minutes of practice, or months? Are there some types of practice that work better than others? How long would it last if you stopped practicing? There is no limit to the questions that can be asked and therefore, it is not possible that all the answers can be known before you come along. This is one of the things that make research fun.

Anorexia and Other Mishaps of Body Schema

Everyone has heard of anorexia. A person, stereotypically a young woman, tries to be thin and in control by starving herself. It's more complicated than that to be sure. For our purposes, it is intriguing that in many cases, the afflicted person sees herself in her mind's eye as fatter than she really is (Figure 3-4). This is an example of an inaccurate body schema. If she were to mentally trace her body contour as you did, she would be way off.

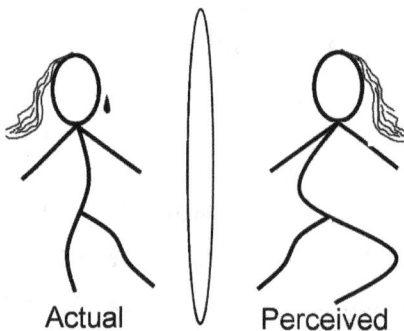

Actual   Perceived

Figure 3-4 The face of anorexia: She looks in the mirror but sees someone huge staring back at her. Original cartoon by Felice Bedford.

It turns out that many normal — I mean non-anorectic — American women who are concerned about their weight have some degree of overestimating body schema. Women shown an array of photos of themsleves in which all but one was digitally altered picked out the wrong one as being the true photograph. The subjects incorrectly picked out a photo that showed them heavier than they really were. If you are interested in practical problems, then discovering the rules governing body schema may allow you design a training procedure to restore body schema to normal in people with anorexia and its less severe variations.

Have you ever been whacked by someone's backpack? That's an example of an incorrect body schema with relatively minor consequences. The offender has not incorporated the backpack on their backs as part of their body schema and so they walk around as if it's not there. Automatic decisions we all make about what spaces we can fit into will be wrong for them, as it's based on a pre-backpack body schema. (This cries out for an experiment to test whether extensions on our backs are harder to incorporate than extensions on our fronts. Why might the former be harder? Good thing baby carriers are worn in front... Backpacks, besides being on the back, just sit there, as some of my students have pointed out.)

Another dramatic inaccuracy concerns a disorder we touched on in the first chapter on perceptions gone awry. If you picked "the man who fell out of bed" then you were given a chapter describing a man who was repeatedly discovered on the floor by his nurses. The chapter is in a well-known book written by Oliver Sachs containing a feast of unusual neurology cases. It turns out this man would fall out of bed because he tried to push a leg off the bed. Except the leg was his own and it was attached to the rest of his body, being his own leg and all, and he would go with the leg over the side of the bed. The man was convinced that the leg was not his and that someone with a terrible sense of humor put someone else's dead leg in the bed with him. It's worth thinking about whether his own leg was not in his body schema. Could we do the reverse, too? Convince someone a leg that isn't his really is? (See Chapter 8 for an answer.)

A body schema with a missing part may be useful in understanding a bizarre affliction where people—believe it or not—hack off a healthy part of themselves. If you happened to see an old television show called "Medical Mysteries", you would have seen a couple of examples. One woman described always pictured herself with no legs below the knees. She was able to show exactly where she pictured each leg missing; it was slightly lower on one leg than the other.

These patients may try to get doctors to amputate and when, of course, they refuse, the patients do something to require amputation. For instance, a limb was placed in a bucket of dry ice for hours to cause irreversible frostbite damage and force the amputation. No mention was made on the show about problems with body schema, but one wonders if somehow, and for some reason, these patients' body schema did not correspond to their actual body. Schizophrenics are the most likely to cut off a body part, sometimes claiming it has been invaded by extra terrestrials or the devil or what have you, but it has also been reported in "normal" people who otherwise are not delusional.

An amputee from the Iraq war who is missing a limb might have the reverse inaccuracy. *Phantom limb* is now well known but was once considered mysterious and had many skeptics. Some people feel like a limb lost to an accident is still there. It even itches, hurts, and has a definite position sense in space. Success with a prosthetic limb is higher if it is started as soon as possible—before body schema catches up and no longer represents the limb. Is a prosthetic limb for an amputee an example of convincing someone that a leg that isn't' their own really is, or is it perhaps more like a rake that becomes an extension of the body?

There are often errors in body schema, most a lot less noticeable. We mentioned earlier that crossed body parts lead to some disruption in knowing which part is where, as do unusual postures. It is often a matter of degree. There are likely numerous small inaccuracies or spots of uncertainty in the body schema of ordinary individuals. They have rarely been explored.

Perceptual Learning Framework

Is anything we have been discussing perceptual learning? Now for the stuff many of you will find boring. We can use the t1-experience-t2 framework and the perceptual learning definition to analyze what's going on. It turns out it is a little tricky both for extensions to body schema with tools and for just movement-induced changes to body schema.

Consider first tools as extensions to body schema, before, during and after experience. Here is an example of the analysis with a hand-held tool, a tennis racquet:

t1: Place the racquet in the hand and stand in a particular position. The actual position of the body is the stimulus at t1. The perception at t1 is the feeling of your body at that instant in time. (Body Schema 1). For instance, the body schema might be as shown in Figure 3-5. Note it does not yet include the tennis racquet.

experience: The experience involves using the racquet repeatedly. The kinds of experience that are required, and how long they should last, are not entirely known. We have seen though that passive holding of an object without doing anything with it is not as effective as using it repeatedly to complete a task, such as to hit balls over a net.

Figure 3-5 The felt body at time 1 (Body schema 1) when first holding a tennis racquet.

t2: Make sure the racquet is in the hand again, exactly as it was for t1, and stand in the same position as at t1. The perception at t2 is where and however your body feels now at t2. If the experience was successful, the body schema should now include the tennis racquet (Figure 3-6).

Figure 3-6 The felt body at time 2 (Body schema 2) after using the racquet.

If the experience was successful at getting the tool to be a part of the body schema then we do have a change in perception from time 1 to time 2 that results from experience. Consequently, a body schema that expands body to include an inanimate object because of extensive interaction with the object is a candidate for perceptual learning. This part is fairly straight forward, although note that the stimulus that gives rise to the perception has to be identical at time 1 and time 2. The pictures that you see that are different at t1 and t2 (Figures 3-5 and 3-6) are not the stimuli, but rather the perceptions. The stimulus at both t1 and t2 is the actual body holding the racquet in the same position, as in the final picture of Figure 3-2. Before experience, the racquet is not part of the body schema but afterwards, it is. The exact same stimulus has to give rise to two different perceptions and when it does so, meets a criterion of perceptual learning.

The tricky part, however, concerns the rest of the definition of perceptual learning. Did this change in perception "fix errors or otherwise improve perceptual systems"? This is a hard question. Was the body schema wrong before experience at t1 because it did not include the racquet? Should the body schema at t2 with the racquet now incorporated be considered to have improved on perception? On the one hand, one could argue that the change from t1 to t2 does not meet this part of the definition of perceptual learning. The argument would be that the perceiver got it right before the experience because a tennis racquet is indeed not part of a human body. Under such a view, the experience created errors, like the damaged brains from Chapter 1 even, rather than fixed them.

But on the other hand, maybe the framework is itself suggesting the situation be viewed differently. Perhaps once you are physically connected to an object, such as when you hold a racquet, it's an error NOT to feel that racquet as part of you. Without trying to sound Zen, two physical things that are connected in some sense do become one. In fact, young infants use connectedness and disconnectedness as cues to inform where one object begins and another ends. Elizabeth Spelke at Harvard showed this. They can do this well before they use other sources of information to determine how many objects there are. If the connected tool has enabled you to perform some action on the world and you rarely leave home without it, it arguably has become part of the body and should be perceived that way.

Now let's consider the framework as applied to a changing body schema from movement. Here, the tricky part comes early because the first analysis that comes to mind is incorrect. It's tempting to analyze it as follows:

t1: Body schema 1, resulting from a position of the body at one instant in time, such as standing with your arm straight at your side.

experience: The experience is moving around because you got a cramp or needed to reach something.

t2: Body schema 2, corresponding to the position of the body after moving at a second instant in time. For instance, the left arm is raised up in the air.

The perception of the body is different at time 1 and time 2 as a result of movement. The change does not count as perceptual learning, however, because the stimulus is different as well. Different stimuli produce different percepts much of the time—that's not learning. The actual body position was different (arm down, arm up) which caused a different body schema (arm down, arm up)—that's ordinary perception, not perceptual learning.

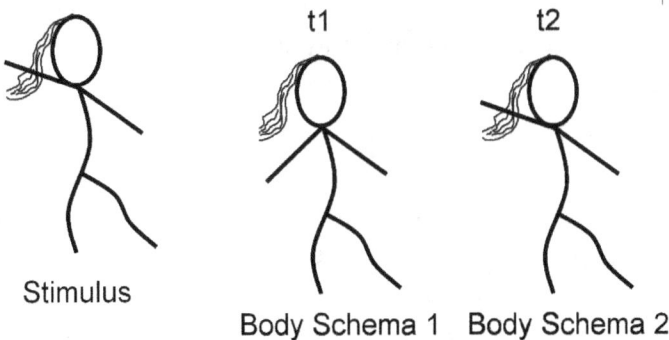

Stimulus
t1   t2
Body Schema 1   Body Schema 2

Figure 3-7 Perceptual learning framework applied to upward movement of the arm. At time 1 (t1), the felt body (Body schema 1) still has the arm down but by time 2 (t2), the felt body has caught up to its correct position (Body schema 2).

But now suppose instead that you moved your body, *but initially did not update your body schema* (Figure 3-7). If you raised your arm up in the air, but did not immediately update your body schema, then both at t1 and t2, the stimulus would be the same. Objectively, your body would have its arm up in the air. But the perceptions would be different at time 1 and time 2)

A candidate for experience between t1 and t2 includes a few seconds spent with your arm in the new position, perhaps much longer for unusual postures. It may also include preparing for an activity in the new position, such as getting ready to throw a rock. Finally, the experience may involve comparing the information about the position of the arm to other sense modalities, such as vision. Cross-modal comparison is useful to create change, as noted in Chapter 2 on development, and discussed later in Chapters 8-11. Experience need not always involve laborious repetition, even though we tend to think of practice that way. Experience comes in all different flavors.

Now it's easier to see that updating the body schema, even just after moving, is a candidate for perceptual learning. There is a change in perception given the same stimulus that results from experience. In addition, the perception that results from experience is closer to the actual stimulus and can be regarded as an improvement, perhaps even removal of an error. If we were not plastic, we could not maintain an accurate body schema and keep track of where we are when our bodies move—this brings us back to what was noted at the beginning of the chapter. Moreover, moving is something that is taken for granted. The interval between the first body schema at t1 and the second body schema at t2 is so short it escapes our notice and we don't realize perceptual learning is involved. Analysis with the framework leads one to the idea that we can catch the body schema unawares in its initial wrong state at t1 following movement, but before being updated. To this end, can we slow down the usually fast time between t1 and t2 so we can see the perceptual learning in action? Imagine being instructed to quickly contort the body in an unusual way while at the same time being required to do several tasks simultaneously, each of which is hollering for your attention. Mental arithmetic and listening to different conversations in each ear would be good choices. If we overwhelm the system enough with unusual postures and mental overload, previously hidden automatic and rapid perceptual learning may be disrupted and slow down sufficiently to witness. Considered along with tool use, we have reached the take-home message early: *We can begin to see that plasticity is needed in adults to accommodate movement rather than only needed in development to accommodate growth.* Experience is involved in the perception of even simple things for adults, such as tool use and moving around.

In fact, the analysis for movement is similar to the analysis for growth that we saw in the last chapter. As applied to growth of the whole body and not just the head:

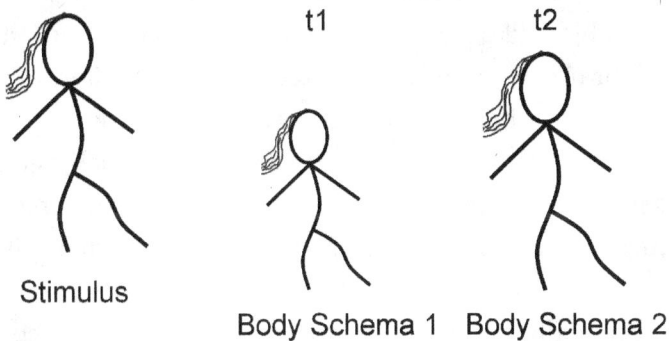

Stimulus

t1   t2

Body Schema 1   Body Schema 2

Figure 3-8 Perceptual learning framework applied to growth. At time 1 (t1), the felt body (Body schema 1) is small, but by time 2 (t2), the felt body has caught up to its correct size (Body schema 2).

Before experience, the body schema is comprised of the small pre-growth body. By t2, the body schema has caught up to the true size if the experience was successful (see Figure 3-8). A candidate for the experience includes again comparison to other sense modalities —for example, comparing felt body positions to seen body positions.

As a reward for making it this far, I end this analysis by introducing a deep issue: Does the mind/brain know or care whether a change to the body is due to growth, to movement, or to extensions? Adding a tennis racquet or swim flippers to our body schema seems like a complex change to the shape of the body envelope, but movements can cause transformations that are pretty complex as well. For instance, imagine curling up in a fetal position and what it does to the shape of the body envelope. Growth can produce complex changes, too, as when the limbs grow more than the head from infancy to adulthood. Thus, the types of transformations may provide little basis for distinguishing among extensions, movement, and growth.

It's Not All Boring

Extensions of the body schema in particular, however, raise something intriguing. By incorporating a tool into the schema, the extensions do not just change the shape of the body schema but they change what is considered *self.* Previously those swim-flippers were an external object like any other. You could contemplate their appearance, their use, whether you wanted them or not, and you could pick them up off the shelf. But after a few hours of snorkeling, they become *your feet.* An object that was external to you, non-self, becomes part of you, self, which is miraculous if you think about it. A piece of plastic you purchased is now more on par with your liver than with acrylic, as if you had transformed into a Borg from *Star Trek.*

A potentially important analogy concerns the concept of self and non-self in medicine. The immune system is responsible for destroying foreign invaders. To do so, it must first be able to determine what is foreign and what belongs there. What's more, this self/non-self distinction in the immune system is also changeable, as it is for body schema. Dramatic examples of this are autoimmune illnesses, where a part of the body becomes incorrectly misclassified as foreign and consequently is attacked mercilessly. Whether there are real commonalties between the issues of self in the immune system and self in body perception remains to be shown, but I believe that there are and that it is an area for fruitful investigation. Chapter 11 is related to this topic.

So, incorporating objects such as tools into the body schema raises the issue of distinguishing between self and non-self. However, this special aspect of adding objects to the body schema does not imply that the mind needs to identify when alternations to the body schema result from such inanimate extensions rather than from movement or growth. Perhaps self and non-self are more relevant to movement and growth than it first seems, even if this component is more obvious when incorporating tools into the schema. I have found too many commonalties between movement and growth[4] to think they are processed separately. Since I'm at least partially committed to the idea that there is no need to distinguish between causes of body schema change, I am driven to be skeptical of any feature that at first glance may appear to distinguish them. You will find, though, that I waver on this issue in later chapters.

While these issues bears on a question of *what* do you consider yourself to be, there is a related question of *where* you do you feel yourself to be. This question is also intriguing and even encompasses topics such as astral travel, also known as astral projection. In astral travel, people report leaving their physical bodies and feel as if they are experiencing the world from somewhere else. Some people claim to do so at will after extensive training and others find themselves accidentally displaced following a severe trauma. Astral travel is not a topic investigated by perception psychologists, but it should be. It has historically been considered supernatural, mystical, and paranormal. There is now a small clinical literature on the *out of body experience*, abbreviated *OBE* for short, which reports on patients with brain damage or drug use who experience being disembodied and able to see themselves. While this research does not reach mainstream literature on normal perceivers, hopefully that is changing. The prestigious journal Science published two studies on inducing OBEs in the laboratory in ordinary individuals[5]. In one study, subjects viewed real-time videos of the back of themselves located a couple of meters in front of where they really were. The subject's back was stroked for one minute.

They watched their backs being stroked in front of them—something normally impossible—at the same time as feeling the touch. This atypical multisensory stimulation caused people to behave as if they were located in front of themselves. Concepts of body schema, sense of self, body part ownership, and whole body experiences, all need to be less marginalized and brought into ordinary everyday perception and how they are altered by learning.

In everyday perception, where we are doing the perceiving from is often more of a mystery than most people realize. Michael Kubovy at the University of Virginia (and the most complex psychologist, perhaps person, I know) coined the term *movable egocenter* to describe that where we are doing the perceiving from has a specific locus and is changeable[6]. Unfortunately, the term, and even the concept, did not catch on. There are also the terms *presence*, referring to where you feel yourself to be, and *telepresence*, referring to moving that location elsewhere but they have had limited use, mostly in the specialized virtual reality and some computer science literatures. When you play a video game and get good at it, you feel as if you are actually there in the screen. Consider also a phenomenon in which perceiving a letter traced on different parts of the body may reveal how we can mentally move ourselves to perceive the world (also a favorite for Kubovy). It is more effective to feel the letters being traced rather than just read about it, but you will not be able do it on yourself, perhaps like not being able to tickle yourself.

The core finding is that if you draw a lower case "b" on a person's forehead, the person says (with eyes closed) that the letter is a "d," but if you draw the same "b" on the back of the head, it feels like it is a "b" instead. It is as if the "person" doing the perceiving normally resides somewhere behind the eyes facing forward, but rather than turning around to read a letter, slides backwards beyond the boundary of the head to reside behind it. Note that in the other out of body experience laboratory experiment published by the journal *Science*, the virtual felt self was also made to move to a position further back behind the real body without turning around. Playing video games, operating machines from a distance, perceiving felt letters, and using imagery to visualize things from other vantage points all exploit the ability to change the location of the "self", a continuum ranging from small, barely noticeable shifts to journeying across the earth through astral travel.

How does someone who claims to astrally travel, such that he is across the room looking at himself, perceive himself? Would a letter drawn on his head, like was done in the demonstration described above, be perceived as if he were located where he actually is or where he says he feels himself to be?

When I was 18 years old, I did a summer internship at the Parapsychology Foundation in New York City, a non-profit foundation that supports the scientific research of psychic phenomena. In my spare time, I read as many of the parapsychology journals that I could from their extensive library collection. One study tried to tease apart these two basic possibilities as applied to ESP, extrasensory perception, by putting an ambiguous letter, like a "b" or a "d", into a sealed box. The researchers wanted to know if the person would mentally read the letter as if he were outside the box looking in or from inside the box looking out. Since ESP is not an easy to replicate phenomenon, the experiment was inconclusive. But what a great idea!

Astral travel is instead an easy to replicate effect that is normal sensory perception, not mysterious extrasensory perception, despite its history as a supernatural occurrence. Letter tracing with ambiguous letters may be a good type of experiment to try for astral travel in order to answer questions of plasticity in location of the self. To my knowledge, this has never has attempted. Hurry up and get motivated by all of this so that you can do it and let me know.

Two Simple Mathematical Equations (Sorry!) are
Important for Perceptual Change

I want to discuss a study that turns out to be much more important than it first appears. The article describing the study is
mercifully only five pages long. I say mercifully because to fully
understand it, I have been telling people to read it at least three
times—and that's if you're one of my colleagues. The article is entitled "Attunement, calibration, and exploration in fast haptic perceptual learning" and it appears in the Journal of Motor Behavior[1]. It
presupposes the reader is well versed in a great deal of technical
terminology and issues from the field of motor behavior. To preview
its importance though, their work includes discussion of two kinds of
error, *variable error* and *constant error*, which, as it turns out, reflect
the two major categories of perceptual learning (the same ones I
introduced in the development chapter).

Have a look first at the task that the authors Jeffrey Wagman and
colleagues used in the study (Figure 4-1). Each subject puts his hand

Figure 4-1 Schematic of a subject
wielding an object without vision.
Figure 1a from Wagman, J.B. et al. (2001)
Attunement, calibration, and exploration in
fast haptic perceptual learning. *Journal of
Motor Behavior, 33*(4), 323–327. Originally
published by Heldref, reprinted with
permission (Taylor & Francis Ltd
http://www.tandf.co.uk/journals)

through a hole in a board, which both blocks view of his hand and restricts movement to just the wrist. The subject now moves his wrist around and tries to figure out how wide or tall the object is that has been placed in his hand.

It is helpful to try the task to see what it feels like. Without the apparatus, just wield an object with eyes closed and be sure to bend only the wrist. Objects I have used are random ones I have lying around, such as a Mini-mag flashlight, the leg from a plastic table, and the aforementioned mechanical reacher. I make sure they are hidden so no one sees how tall or wide they are. This is difficult to do on your own because you would have to see the object first and that would give away its dimensions. If you cannot easily get a partner to test one another, you could try setting aside many objects and then picking one up at random with your eyes closed. Wield the object using wrist motion alone and see if you can tell how tall the object is. When you think you know, open your eyes, but without looking at the object, mark two endpoints on a surface that correspond to the height of the object. When I have tried this on others, performance varies greatly. I have had some people who were so close to perfect it looked like we had staged the whole event. Others were not quite as perfect, but still, no one was ever so far off as to cause any laughter. It is not a task you usually attempt every day, but if you try it, you may surprise yourself at how good you are. You also don't know how far off you would be, so go ahead and give it a try.

Figure 4-2 shows the hand exploration pattern of one subject from the published experiment. Each dot shows where in 3-dimensional space the subject placed his hand at one instant in time while trying to figure out an object's size. As you can see, the wrist is moved all around to try to figure it out. I am often surprised that when I discuss this with students, not all of them know what 3-dimensional space means. If it is not a familiar term for you either, consider the space around you: the world you look at, reach in, and move around in. Move your arm left to right and that's one dimension of space (often called X). You can also move it up and down (Y), and you can move it

in and out (Z). Each dimension is at a right angle—90°—to each of
the other two. Three of these independent dimensions that are 90 °
from one another describe the space we inhabit. It's one of the
properties of our world that is so obvious in our everyday lives that
we take it for granted. Before perception psychologist Roger Shepard
at Stanford retired, he noted that because people take 3-D space for
granted, they say things like "I wish I had a bigger room for my
furniture", but never "I wish I had a 4-dimensional office so as to have
more degrees of freedom for arranging them." [2]

Figure 4-2 Each dot is a
location of the hand
during a single bout of
object wielding in the
object exploration study.
The first location is
indicted with a 1, and
every 100th location is
indicated with its number.
Figure 1b from: Wagman, J.B. et
al. (2001) Attunement,
calibration, and exploration in
fast haptic perceptual learning.
Journal of Motor Behavior,
33(4), 323–327. Originally
published by Heldref, reprinted
with permission (Taylor &
Francis Ltd http://www.tandf.co.uk/journals)

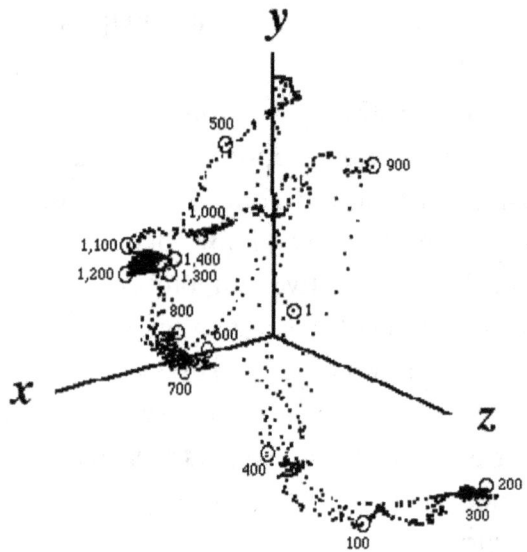

Since the article is difficult, I will go through it step by step. The
authors note a distinction others have made in the literature between
fast and slow perceptual learning. Wagman and colleagues were
interested in exploring the properties of fast perceptual learning in
which it takes relatively few trials to get improvement. The task they
chose was the one introduced above. Each subject wields a tool in his
or her hand and estimates its height or width without using vision
(*haptic exploration*). Note that because of a lifetime of using small
hand-held objects, as we encountered in the last chapter on incorporating tools into the body schema, their expectation was that people

would already be decent at the estimations. It was also expected that the subjects would not be perfect, however, because we are rarely asked to make such precise judgments explicitly. They discuss two basic types of errors one could make on this task, which they characterize in equation form, Constant Error (CE) and Variable Error (VE). Next, they have a prediction that feedback will have different effects on these errors. They suggest that feedback on how well a person has performed on a judgment is needed to get rid of CE, but that feedback is not needed to get rid of VE. (They refer to feedback as *knowledge of results*, or KR, in keeping with the motor literature.) The researchers therefore sought to determine if there would be improvement on the haptic exploration task with practice and additionally, to assess the role of feedback on the two types of improvement.

Now read the procedure they used:

"We demonstrated the use of the report apparatus before the experiment with an object....On each trial, participants wield an object with their right hand (as depicted in Figure 1a) [*note this is our Figure 4-1]* for as long as they needed to reach a judgment. The participants could adjust the report apparatus continually while wielding the object. When satisfied the participant notified the experimenter, who terminated the trial. Except for practice with KR, participants were given no explicit knowledge of the size range of the objects."

So subjects could take as much time as they needed to make the judgment, which initially was an average of 48 seconds. The way subjects gave their answer was not verbally—which would be hopeless—but by physically moving two markers on the apparatus to correspond to the height or the width. Half of the subjects were asked to match the height of the wielded object and the other half of the subjects, the width. We can extract the structure of their experiment as follows:

Day 1 18 trials   Tested, stimulus set A   No feedback
Day 1 25 trials   Practice, stimulus set B   ½ subjects get feedback
Day 2 25 trials   Practice, stimulus set B   ½ subjects get feedback
Day 2 18 trials   Tested, stimulus set A   No feedback

The 18 test trials on both days consisted of 6 stimuli of different heights and widths, all the same basic shape and design as depicted above in Figure 4-1. Each stimulus was repeated 3 times in random order. The 25 practice trials on both days consisted of 5 additional stimuli with different heights and widths, each repeated 5 times. Note how subjects are tested both before and after practice, and in this case, practice and test were split across 2 days. If you are wondering how they were provided with feedback, it was through vision. Following a judgment, those subjects assigned to receive feedback during the practice trials (half of the subjects) were now shown the object. It was placed right next to the reporting apparatus where the exact width or height of the object could be visually compared to the width or height between the two markers that they themselves had set. During testing, none of the subjects received feedback (since it was a test...)

Before getting to the results, let's first consider in detail the two types of errors that people can make.

Variable Error and Constant Error

The equations are very simple, even if it doesn't look that way. If you can put aside any knee-jerk math aversion reaction you have, you will be able to use them. More important than the calculations is what each error means psychologically.

Formulas for the two main errors, constant error (CE) and variable error (VE)

$$CE = \frac{\sum (Xi - D)}{n}$$

(4.1)

$$VE = \sqrt{\frac{(Xi - M)^2}{n}} \qquad (4.2)$$

where: Xi = perceived value on Trial i
D = the actual correct value
n = number of trials
M = subject's average judgment (over the n trials)

A shortcut for calculating constant error
    CE = $M - D$ (4.3)

where: M is the mean, $M = \dfrac{\sum Xi}{n}$

Formula for another error, Absolute Error (AE)

$$AE = \frac{\sum |Xi - D|}{n} \qquad (4.4)$$

Here are three examples of how to calculate the errors for Bob, Jack, and Lila, followed by what the patterns of errors actually mean. Suppose that one of the objects explored by a subject, Bob, is 5" inches wide. Suppose further that four of Bob's judgments for that object in an experiment like the Wagman study were 3, 5, 10, and 2". I know a Bob and he jumps all over the place in an argument; he's often wrong but thinks he is right. Thinking of this subject as Bob is a good way for me to keep track of his data. Looking at the four data points, note his first estimate was a little narrower than the actual width, then he was dead on, but then he was way off and too wide, and finally the fourth judgment was back to underestimating by 3". Bob is clearly a person that is making mistakes—errors. The formulas allow you calculate how much error and of what kind. We will not be spending much time on absolute error (AE), which was also considered by Wagman et al., but let's calculate that first anyway. You can think of it as an overall measure of error that has features in common with the other two more specific kinds of errors.

If you are happy using formulas, just calculate the result. Here is a step-by-step guide for those who prefer one, though it may make it seem more complicated than it really is. From the formula for AE (equation 4.4), take first the judgment on the first trial, which is 3, and subtract from that the correct value of the width, which is 5. That leaves a value of -2. Then take the absolute value of that number, which is +2. (If you have forgotten elementary math, absolute value converts negative numbers to positive ones, and leaves positive numbers unchanged. It's represented by the notation $|\ |$. Take the absolute value of anything within the straight brackets). Repeat this same procedure for every trial. Thus, on the second trial, the judgment is 5; subtract again the actual value of the stimulus, 5, which leaves 0 and take the absolute value, which remains 0. The third trial is 10, subtract 5 and take the absolute value, which leaves 5. Finally, take the last trial, 2, subtract the actual value 5, which leaves -3 (negative 3)and take the absolute value, for a result of 3. Next, as seen in the equation, sum (Σ) those 4 numbers, which add up to 10. Then divide that sum (10) by the number of trials, namely, 4. This leaves a value of 2.5, which is the value of the absolute error:

$$AE = \frac{2 + 0 + 5 + 3}{4} = \frac{10}{4} = 2.5$$

For constant error (CE), to do it the long way (equation 4.1), you would again take Bob's judgment on each trial and subtract the correct value for the object as you did for absolute error. For this error, however, do not take the absolute value of each result. Instead, just sum those 4 results for the 4 trials the way they are. In this example, they sum to 0. To get the constant error, divide this sum by the number of trials, as you did for absolute error:

$$CE = \frac{(-2) + 0 + 5 + (-3)}{4} = \frac{0}{4} = 0 \quad \textit{shortcut:} \quad CE = 5\text{-}5 = 0$$

The equation for the shortcut is also shown above (Equation 4.3). Take the subject's average judgment of all of the trials first and then subtract the correct size of the object just once. The average of the 4 judgments (3, 5, 10, and 2) is 5, and subtracting the correct size, which is also 5, leaves a value of 0. Whether calculated by the long or short method, the constant error in this example is 0. Although Bob makes errors on most of the trials, he has no constant error. We will come back to this soon.

To calculate variable error, first calculate the mean of all the subject's judgments, which was calculated above to be 5. Then, from each individual judgment, subtract that mean and square the resulting number. As with the calculations for the previous errors, add the numbers from all of the trials (judgments) together and divide again by the number of trials. Finally, take the square root and the result is the variable error. In this example, Bob's first indicated the answer was 3; subtracting 5 leaves -2, square that and you have 4. The second judgment is 5, subtracting 5 leaves 0, and 0 squared is 0. The third is 10, subtract 5 again, which leaves 5. Square the result, which is 25. Finally, the 4th trial is 2; subtract 5 which leaves -3, square that number for a result of 9. The sum of these 4 calculations is 38. Divide by the total number of judgments, 4, yielding 9.5. Finally, taking the square root gives the variable error, a value of 3.08.

$$VE = \sqrt{(4+0+25+9)/4} = \sqrt{38/4} = \sqrt{9.5} = 3.08$$

Think for a moment about the variable error. Since you are subtracting the subject's own mean from each judgment, you are getting a measure on each judgment of how far off they are from their own average performance. Why take the square of the number? Even equations should be questioned. If you didn't take the square then overestimations and underestimations would cancel each other out. For instance, if a subject underestimates by 3 inches (-3) on one trial, and overestimates by 3 inches on another trial (+3), the positive and negative numbers would sum to zero if the square of each number wasn't taken first. This would give the false impression that there was no error, when, in fact, the subject was highly variable. By taking the square of the number, the result is always a positive number and over and underestimations will not cancel each other out. The square root is taken at the end to bring the scale back to the scale of the original numbers before they were squared. Things like being told to take the square and square root are not so mysterious when you think about them.

Consider another example and calculate the absolute, constant, and variable errors. To preview this example, the absolute error is identical to the previous situation, yet the component errors are different. This subject, Jack, shows more of a constant error but less of a variable error than the previous subject does. His answers for the same 5" stimulus were 8, 8, 7, and 7. Be careful with the calculation because in this example, M ≠ D; that is, Jack's mean judgment is not the same as the correct value of the stimulus, unlike Bob.

For absolute error, subtract the correct value of the object, 5, from each item, take the absolute value, sum, and divide by the number of trials:

$$AE = \frac{3 + 3 + 2 + 2}{4} = \frac{10}{4} = 2.5$$

For constant error, compute the average of 8, 8, 7 and 7, which is 7.5, and subtract the correct value of 5:

$$CE = 7.5 - 5 = 2.5 \quad \text{or the long way:} \quad \frac{3 + 3 + 2 + 2}{4} = 2.5$$

Finally, calculate the variable error. Subtract Jack's mean, 7.5, from the first judgment of 8, which leaves .5. Squaring that number gives .25. Repeat the process for each of the remaining 3 trials, sum them, divide by the total number, and finally take the square root.

$$VE = \sqrt{(.25 + .25 + .25 + .25)/4} = \sqrt{1/4} = \sqrt{.25} = .5$$

Jack's variable error is smaller than Bob's. The final example, Lila, is even more extreme because her variable error is zero. She always said the 5" stimulus was 8" on all 4 of the trials. For Lila:

$$AE = \frac{3 + 3 + 3 + 3}{4} = \frac{12}{4} = 3$$

$$CE = 8 - 5 = 3 \quad or \quad \frac{3 + 3 + 3 + 3}{4} = 3$$

$$VE = VE = \sqrt{(0 + 0 + 0 + 0)/4} = 0$$

I have never met a student in any of my classes, whether they hate math or love it, are male or female, or are good at it or bad at it, be unable to calculate these errors. This is provided that anyone that does not remember basic principles is shown step by step how to perform them. Granted, success is not too surprising given that only addition, subtraction and division are involved, but nonetheless, mathematics performance has a bad reputation and it need not. I fear I have contributed to this since calculation details in a non-interactive medium are droll. My apologies for showing all the steps. Much harder for students, rather ironically given the frequent aversion to simple math, is thinking about how to interpret the numbers they have just successfully calculated.

In the first example, the constant error was zero, but Bob had a variable error. In the third example, the situation was reversed: Lila had a constant error, but did not have a variable error. For Bob, note how his judgments averaged perfectly to the real value of the stimulus (5), yet the individual ones (3, 5, 10, 2) were all over the place. In contrast, note how Lila's average is way off from reality (8), yet she was very consistently way off each time (8, 8, 8, 8). I picked Lila for this example because always overshooting the target can be pretty comical and the Lila I know is very funny. You may have encountered these different patterns of errors in any kind of sport requiring hand-eye coordination because they are particularly noticeable there. Suppose you were throwing darts, shooting a rifle, or even bowling. You may know someone, perhaps yourself, who always "pulls to the right". If so, you might need to intentionally compensate and aim further left or else you may always miss the target by being too far to the right. Such a person is like Lila—he or she is consistent and predictable, yet always has a *constant error.*

Alternatively, other people may not have such a systematic bias, yet still make errors because they are not being very consistent in their performance. On average they may be accurate, but one time they may shoot too far right, and another time too far left. Such people are, in the extreme, like Bob—he or she has no constant error, but does have a *variable error.* People who are new to a complex task often show this high variability. Or watch children if you want to see large variable errors. Children are new to all tasks. Xander, the 5-year-old son of a friend, wanted to take his turn filling a bucket that was 10-feet away with a garden hose (don't ask). The task was very easy for the adults who all honed in rapidly on the right location to direct the stream of water. Xander sometimes hit the bucket, but the stream moved in a very wide arc swinging back and forth, sometimes going too far to the left and other times missing too far to the right. Fortunately, getting soaked is always a good thing in the blazing Tucson sun. A mom of a younger boy told me this was exactly what it like as her son took aim learning to use the facilities…If you have taken statistics, you can think of the variable error as the standard deviation and the constant error as the mean difference. Perfect performance on a task would require having neither error by removing both the constant error and the variability.

The distinction between constant error and variable error is very important. The two errors reflect the two major kinds of perceptual learning. Getting rid of variable error is what the process of *perceptual expertise* is about, and getting rid of constant error is what *perceptual adaptation* does.

More on the Fast Haptic Perceptual Learning Experiment

Going back to the Wagman and colleague article, we can now finally consider the results of their experiment. Recall they predicted that before any special practice, subjects would be decent, but not perfect, at estimating the dimensions of an unseen wielded object. They also predicted that practice would have different effects on the two kinds of errors: Just practice with wielding and estimating, but without any feedback, would be sufficient to reduce variable error, whereas feedback would be needed in order to reduce constant error.

The results confirmed their predictions. CE decreased only in the feedback group. VE decreased whether subjects got feedback or not, and did so equally in the feedback and non-feedback groups. Thus, it was just practice with the estimations that made subjects less variable in their estimates of height and width. Getting feedback on the accuracy of their estimations then reduced any constant biases subjects had if they initially over or underestimated the objects' dimensions. The results are consistent with some things we will take note of later when looking at each perceptual learning process, perceptual adaptation and perceptual expertise. That is, researchers in the field of perceptual adaptation find that some form of feedback is crucial for learning, whereas investigators of perceptual expertise (AKA perceptual discrimination learning) often report that mere repetition is sufficient for learning. But there is time for that later.

A detail of the experiment results—perhaps the last thing you want to see right now—is that overall, the variable error was higher for height estimates than for width estimates. There could be intrinsic differences in how people perceive height and width, but the particular stimuli used in the experiment may have influenced this result. Inspect the actual measurements of the objects used in the study.

Can you figure out why the stimuli could produce differences between subjects' height and width estimates? Do you notice any
difference between the height and width numbers? Try looking at

Table 4-1 Object Sizes in the Object Exploration Study

Test Objects					Practice Objects				
Width	Height	Mass	l_1	l_3	Width	Height	Mass	l_1	l_3
18.1	7.6	391	19.3	.47	10.2	10.2	393	13.1	.06
13.3	8.9	392	15.5	.55	11.8	12.7	704	28.6	1.9
20.3	10.8	881	50.1	1.85	12.6	14.0	914	38.9	3.0
16.5	16.5	1674	83.2	7.54	9.5	15.9	889	31.0	3.5
10.2	21.0	1660	62.8	11.46	27.9	12.7	1665	132.9	4.9

Width and height in centimeters, mass in grams, l_1 and l_3 in gem^2 Adapted from Table 1
Wagman, J.B. et al. (2001) Attunement, calibration, and exploration in fast haptic
perceptual learning. *Journal of Motor Behavior, 33*(4), 323–327. Originally published by
Heldref, reprinted with permission (Taylor & Francis Ltd http://www.tandf.co.uk/journals)

the dimensions of the objects used to give subjects practice and how
those compare to the dimensions of the objects they used to test
subjects. For width, the practice objects ranged from 9.5 to 27.9 cm
and the test objects ranged from 10.2 to 20.3 cm. For height, the
practice objects ranged from 10.2 to 15.9 cm and the test objects
from 7.6 to 21 cm. If that still just looks like a bunch of numbers, note
that for width, all the test item dimensions are contained within
those of the practice items. Subjects get practice on a wide ranging
set of widths and all stimuli they then are tested on are within that
range. Contrast that with the objects used for height. Here, subjects
are tested on items with sizes that fall outside the range they practiced on. For instance, when the test object is 21 cm tall, it is taller
than any of the practice objects encountered. If the variable error is
high for height specifically after practice (it was not reported separately before and after practice), then one potential cause is that the
practice stimuli did not generalize as well to the test stimuli for
height as they did for width because they were not as similar. There
are other factors that you can explore about the data set that may also
have been influential for any differences found between width and
height, even before practice. For instance, how does just the test
range of stimuli for height compare to width before any practice? For

how many stimuli is the object's width larger than its height? Is this the same for practice and test objects? Might it matter? I have no particular reason for believing any of these did influence the results but you often do not know which details make a difference until someone tries to replicate the study. It is good to analyze as many details as you can.

I singled out the issue of generalizing from the practice items to the test items to point out that the Wagman and colleagues study on manually exploring objects is really a study on generalization, even though it is not described as such. The main conclusion of the experiment is that feedback on practice is only needed to reduce constant error but not variable error. However, if we are being really careful and literal, the study does not show that. Instead, the experiment shows something like: Feedback is needed to *generalize* any removal of constant error to new stimuli. This is because subjects were never actually directly given practice or feedback on the exact stimuli they were tested on. Testing on stimuli different than the practice stimuli has some advantages—like we know subjects' improvement with practice cannot be because they just memorized the height or width of the practice stimuli—but it also has some drawbacks. I realize this is confusing. Chances are that the original conclusion will turn out to be correct. Still, one should keep those nagging, often complex, issues in the back of one's mind in the event that they may become relevant at a future time.

Since we are on this general topic of the details of the experimental procedure, we can also extend discussion to ask what variations of the experiment it might also be useful to conduct. Recall we did this for body schema in Chapter 3 as well. One issue is whether the result would be any different if the feedback were given in some way other than through vision. We could also ask whether more practice would reduce the errors yet further or even eliminate them completely. To relate the experiment to the last chapter, were the wielded tools here incorporated into body schema and how might one test for this? Think of the variants you would want to explore.

To continue dissection of the experiment, take a look at Table 4-1 again and note there are columns labeled "mass", "l1" and "l3". These refer to the information that is available to indicate the width or height of an object being manipulated but not seen. Remember that ice cream cones and pens and anything else you might grab a hold of do not have muscles, skin, tendons, or receptors that would indicate where the object begins and ends. (If you are thinking about the turkey drumstick at Thanksgiving, ok, it has muscles, but they are not attached to your brain so it doesn't do you much good as a source of information.) How then can people figure out what the dimensions are of a hand-held object?

Golf players advanced enough to prefer some types of clubs over others may have searched for clubs with a high *MOI.* Such a golf club is desirable because it would be very forgiving and not turn much if you hit with it off-center. MOI stands for moment of inertia, which refers to resistance to rotation (with respect to an axis). This just means that when you move your wrist around while holding a tool, the resistance you feel is a detectable piece of information in your own body from which you can extract the boundaries of the external inanimate object you are grabbing. Weight, shape, and symmetry of the object will influence the MOI around each of the three dimensions of space that you move your wrist. In the table, l1 and l3 refer to components of MOI. Mass (weight) of an object by the way is ambiguous with respect to revealing to you its dimensions. Even assuming each object is made from the same weight material, a heavier object could indicate it has greater width or it could mean it is has greater height.

One final note about the article is that it remains to be seen whether the distinction drawn in the motor literature between fast and slow perceptual learning, that the authors also made use of, is a meaningful one (See Chapter 12 for discussion of the time course of perceptual learning). Similarly, the importance of other concepts they discussed, such as *attunement* and absolute error, is not clear. Indeed, it is not even clear whether every improvement found in this task would survive the scrutiny of our perceptual learning framework. But it is certain that the distinction in perceptual learning reflected by variable error and constant error is real. The importance of this article lies in its invoking constant error and variable error with their simple mathematical formulations in relation to changes in perception with practice.

A Little More on Variable and Constant Error

It is rare for the connection we have seen between statistical errors and changes in perception to be made. One exception occurred in the 1950s by Eleanor Gibson[3] (see Figure 4-3). E. Gibson—to distinguish her from husband JJ Gibson of general perception fame—put the perceptual learning process of perceptual expertise on the map, as we will see in Chapter 6.

I had the pleasure of meeting "Jackie" Gibson in the 1980s when I was a graduate student and she came to visit the University of Pennsylvania. Alas, by then my scrutinizing questions were met solely with "read my husband's book" and no further insights could be obtained. I was told—by the aforementioned "do not die" ex-mentor Henry Gleitman—I should treat her with the kindness one gives one's grandmother. As a graduate student, however, I did not agree and believed that in the context of academia, a person should be judged and treated based solely on his or her current academic merit. You can decide for yourself how you would behave. Nonetheless, her writings remain influential, interesting, and too often, the details lost or misunderstood.

Now go back and consider Chapter 2 on growth and sound localization. Recall, we saw that there were two kinds of errors there as well. We discussed that two processes in development already mirrored the two major processes of perceptual learning in general. One of them corresponds to perceptual adaptation and the reduction of constant error. The other corresponds to perceptual expertise andt he reduction of variable error. Which is which? This is a good time to take a break again and think about it if it's not immediately obvious.

REDUCTION OF VARIABLE ERROR

Ordered
Response R_1 R_2 R_3 R_4
Categories

Stimulus
Continuum

REDUCTION OF CONSTANT ERROR

Ordered R_1 R_2 R_3 R_4
Response
Categories

Stimulus
Continuum

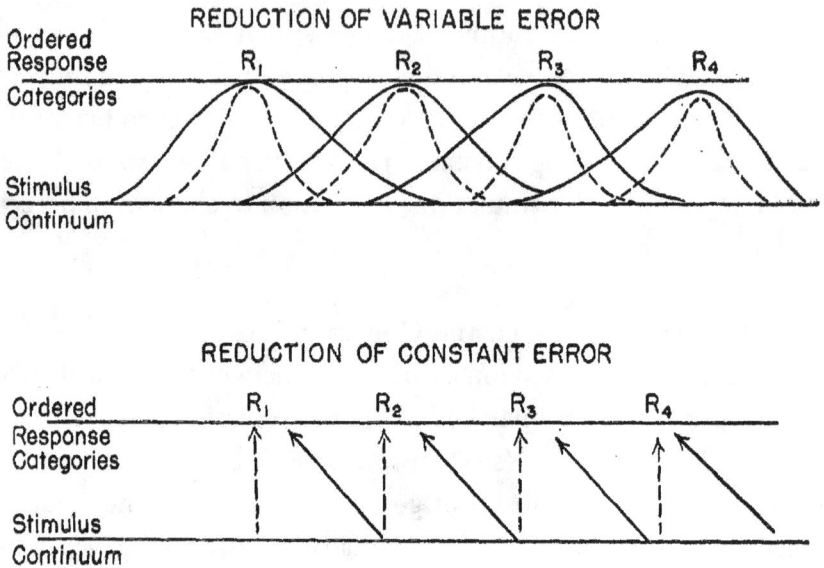

Figure 4-3 Distinction between variable and constant error and its relation to perceptual learning by Eleanor Gibson. In the top panel, solid curves show range of stimulation that elicit a given response before training and dotted curves show the narrower range after perceptual expertise—then called differentiation—training. In the bottom panel, solid lines show the relation between a stimulus and response and dotted lines show the relation when the constant error has been eliminated. Gibson, E. J. (1953). Improvement in perceptual judgments as a function of controlled practice or training. Psychological Bulletin, 50(6), 401–431. Published by the American Psychological Association (APA); reprinted with permission.

Another important tip for everyone is to always integrate material from one topic to the next. I learned this, too, from my students through their surprise when previous topics are relevant to current ones. Even my best-scoring students will often completely purge a section from mental storage once we have moved on to a seemingly new topic, falsely believing it will never again be needed. Everything will be needed again. The link to development is that the potential error produced by uncorrected body growth corresponds to constant error. Change in head size would produce a systematic predictable error in sound localization were it not eliminated with perceptual adaptation. In addition, we also discussed how with age, children get more precise in their perceptual judgments. This process corresponds to the reduction of variable error through perceptual expertise.

If you like testing yourself, here is a question on calculating errors and interpreting their meaning. The answer is in the Notes section at the end of the book. You hear about an experiment on face perception. (In the next chapters, face perception will be addressed.) People are shown photos of the nose and upper lip for a variety of faces. Each photo is shown for 3 seconds and is repeated a total of 4 times. The subjects' task is to reproduce the distance between the nose and the upper lip using a ruler. One of the photos measures exactly 1 inch between the nose and upper lip. Suppose you observe the following data for this 1″ distance on each of the 4 repetitions: 1.5, 1.0, 0.9, and 1.4″. Furthermore, after practice/training, the reproductions for the same 1″stimulus are 1.2, 1.2, 1.2, and 1.2. What are the absolute error, constant error, and the variable error for the data both before and after training? Do you think the subjects received feedback as part of training? Explain why or why not. Here is a hint: You should be able to answer the second part even if you are not inclined to calculate the numbers asked in the first part[4].

Where Do We Stand Now?

In the last two chapters, we have started to see the need for perceptual learning in adults. Changes in perception are needed to update body schema both to incorporate tools and to remain accurate while we move our bodies through space (Chapter 3). Changes in perception, such as haptic perception, are needed for improved accuracy when a task requires greater precision (Chapter 4). Chapter 2 explained the need for perceptual change during development if perception is to remain accurate with physical growth. And even Chapter 1 implied a use for perceptual learning to accommodate sudden unexpected damage to any part of a perceptual pathway. But it is the two chapters on normal adults that point to a need for perceptual learning on an ongoing basis. Perceptual learning does not result from a mechanism that is simply initiated by development or damage and then turns off when the body is fully grown or restored to health. Rather, we can already see that it is an integrated part of perception that adjusts to movement and conforms to any task for which we need better perception.

Consider what appears at first to be a local question on the particular haptic perception task we have been discussing but is really a very general question. Would you ever need to determine the height or width of an object wielded in your hand? Sure. To use a tool effectively, knowing the boundaries of the object is critical. Although we would rarely have to explicitly reproduce an object's dimensions on a ruler, we would need to implicitly know its dimensions in order to effectively hit a ball with a bat or pluck the grapefruit with the picker. When you use your own hands to manipulate objects, that information is provided by the sensory receptors in the muscles and joints. Perceiving where your fingertips end then allows effective manipulation of objects with those fingertips. As noted several times, however, sensory receptors on inanimate objects are simply not an option for perceiving where those objects end. The Wagman and colleagues article suggests how the needed information is obtainable for objects that are not a part of us—moving them around provides different resistance in our hands depending on how tall and wide they are. Detecting the information would then allow effective manipulation of the world with tools as well as fingertips (and perhaps the incorporation of the tools into the body schema).

But here's the deep question that lurks beneath. If perceiving the dimensions of a hand-held object is important for effective tool use and the information to do so is available (through resistance to rotation, moments of inertia) *and* we have had experience manipulating objects in our lives, then why was any training required is this experiment? Why do perceptual judgments become more accurate with further practice—or any practice? The answer is: *Perception is lazy.* Perception is only as accurate as you need it to be for the task at hand. It is one of the reasons perceptual learning is needed on an ongoing basis. Perceptual learning is needed to mold perception to tasks; much of the time, it provides a way to improve as we need perception to improve and to diminish when we do not. Why in turn that should be is a theoretical question best addressed after we have seen more.

And there is more to see. The next several chapters delve into the two major types of perceptual learning glimpsed in the first few chapters. First up is perceptual expertise. Chapters 5, 6, and 7 explore how practice leads to an improvement in perceptual precision which results in an ability to actually see things that you never could see before! Chapter 8 sets the stage, in a few places almost literally, for perceptual adaptation. Then Chapters 9, 10, and 11 show how this process has the potential to alter perception of the very fabric of our existence that you take for granted: space, time, and well-being.

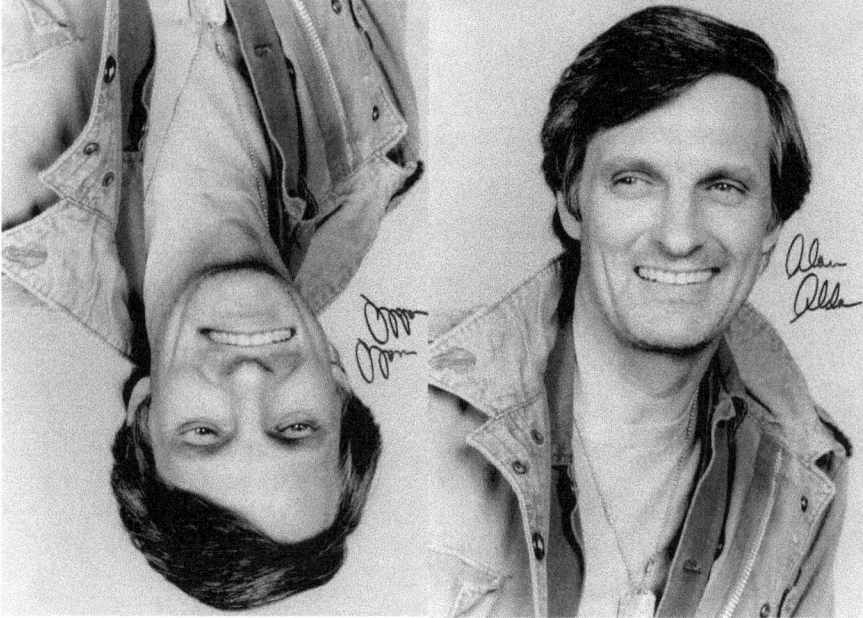

Freaky faces. **Figure 5-11** (top) hides a large head. Do you see it? **Figure 5-16** inverts actor Alan Alda. Turn the book upside down for a startling face! See text. Top: *The Slave Market with Disappearing Bust of Voltaire*, by Dali, 1940 © Salvador Dalí, Fundació Gala-Salvador Dalí/Artists Rights Society (ARS); reprinted with permission. Bottom: publicity photo of Alda was *thatcherized* by Laura P. Garcia and Felice Bedford.

5

Do They All Look Alike? An Introduction to Perceiving Faces from Other Races

On the way home from a trip on a freezing cold day in March, I called ahead for a taxi to meet me at the train station. When the driver paused to swig some cough syrup, I got a good look at him in the mirror and asked: "Hey, didn't I have you on the way over here?" He said: "No, No, it wasn't me, but ya know, all us cab drivers look alike" and proceeded to chuckle and cough in alternation. One takes the bad with the good from taxicabs, the flu bugs and the insights.[1]

But I really want to begin this chapter with a task. I'd like you to

Figure 5-1 Training figure. Study the form for a few seconds and try to remember it. *Cover it before looking at the test figures.*

look at a form (Figure 5-1) and try to remember it since I'm going to ask you to pick it out of a line-up. Well, a psychologist's line-up. Do you think it's "A" (left) or "B" (right) in the test picture (Figure 5-2)?

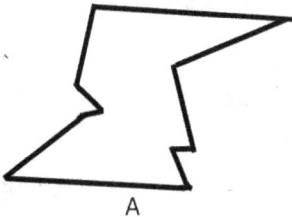

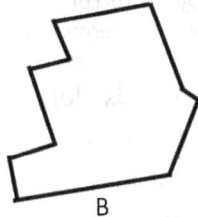

A

B

Figure 5-2 Which figure have you seen before? Avoid looking at the previous pic.

"A", you say? You are right. Everyone gets it right. Now try this one:

Figure 5-3 Second training figure. Study it for a few seconds and try to remember it. *Then cover it.*

Do you think it's on the left or the right"? (Figure 5-4.) It's harder.

Figure 5-4 Test figures for Figure 5-3. Which is the one you have seen before? Avoid looking at the training figure until you decide.

The correct answer is the shape on the left. And finally, try one more:

Figure 5-5 Final training figure. Study it for a few seconds and try to remember it.

Try to study it. Ok, is it on the left or the right? (Figure 5-6.)

Figure 5-6 Test figures for Figure 5-5. Which is the one you have seen before? Figures 5-1 to 5-6 adapted from Rock, I., Halper, F. & Clayton, T. (1972). The perception and recognition of complex figures. *Cognitive Psychology, 3,* 655-673. Published by Elsevier; reprinted with permission.

The correct answer is "left" again. This one is easier than the one before it. In fact, this is what it feels like to recognize members of your own race. The task before is what it feels like to try to recognize members of a different race. Let me point out something quickly about the figures. The line segment that you just got right is the exact same line segment from the bigger figure that you may have gotten wrong. To verify this, look at the left choice of Figure 5-4 and note that the training figure from Figure 5-5 comes from its upper left side. If you are not completely convinced, cut out the line segment or fold it over until you see it's the identical segment from the bigger closed shape. To preview: *It's not as if you can't process information, you often just don't know where to look.*

Ok, now let me go to a proper beginning after two false starts: THEY ALL LOOK ALIKE! This is an experience that people have when looking at individuals from a race different than their own. People think it, even if they do not say it out loud. Is this just folklore that we think this? No. It's a scientific fact. Is it racist to have that experience? No.

Look at a graph from an experiment on nationality and face perception conducted by psychologist Nathan Doty (Figure 5-7).[2] It

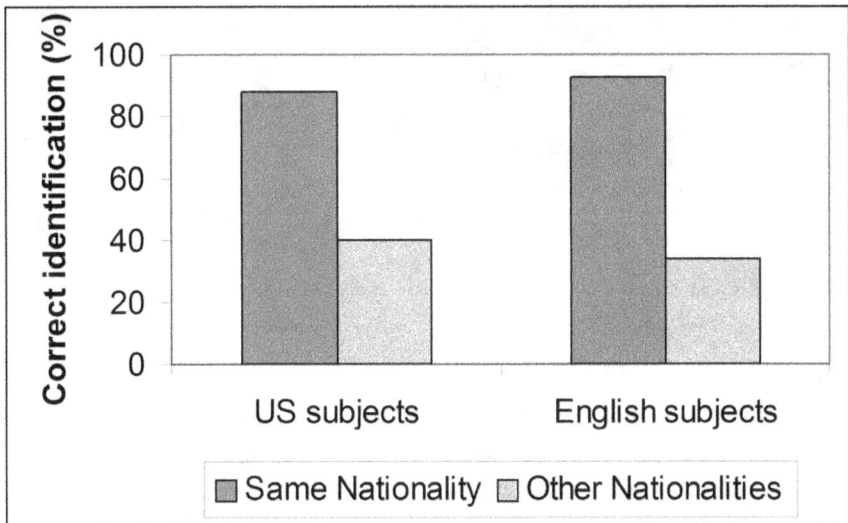

Figure 5-7 The effects of nationality on face recognition by U.S. and English subjects. Data extracted from the Doty article.

shows that subjects from the United States recognized faces of other Americans they just "met" 88% of the time, yet could only recognize those from other countries, such as Belize, England, and France, 40% of the time. On the other hand, subjects from England could recognize the individuals from England (even when the Americans could not), but did much more poorly on people from other countries.

The procedure he used is representative of a number of studies in this area. A photo of a face in front view is presented very briefly—just 1/10th of a second. Subjects are then shown a selection of faces that are all side views and they try to pick out the individual seen before. In this experiment, 10 faces were shown and the chance of being right was 10% if a subject were merely guessing. Despite seeing a face for such a short time, the subjects are doing better than chance —though twice as good on their own countrymen. The article does not show the faces that subjects saw but in Figure. 5-8 you can get a sense of what it's like to look at faces with hair and other identifying features stripped away.

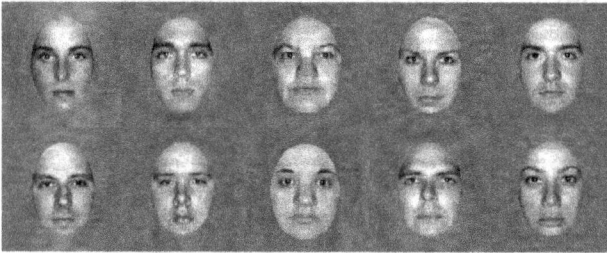

Figure 5-8 Front view of faces without hair, hats, facial hair or jewelry. How easy is it for you to distinguish one face from another? Reprinted with permission of Macmillan Publishers LTD: NATURE Gold, J., Bennett, P. J., & Sekuler, A.B. Signal but not noise changes with perceptual learning. *Nature, 402*, 176-178, copyright 1999.

The races of the photos and the subjects varied in Doty's experiment but were mixed in with the country of origin. His study is actually showing that people of other nationalities all look alike, independent of race. You get comparable data if you just use race, the more socially controversial finding. Caucasian Americans find African American faces and Asian faces difficult to distinguish. African Americans have more difficulty with Caucasian and Asian faces than they do other African American faces. Chinese Americans have difficulty with Caucasian and African American faces. These have all been shown by carefully controlled laboratory experiments like the experiment on nationality.

Cross-racial perception problems have also been verified outside the sterile laboratory setting. In one experiment[3], convenience store clerks in Texas were visited by three customers, one African American, one (Anglo, non-Hispanic) Caucasian, and one Mexican-American man. Except the customers weren't customers at all, but confederates assisting the researchers. Each "customer" had a set script to follow, including engaging in a task that would be reasonably memorable. For instance, one of the tasks was paying for a 99-cent item all in pennies while apologizing for the trouble. The same day that a clerk was visited by the customers, two additional confederates came by acting as law interns. "There's reason to believe that three men who are persons of interest may have visited your store", they said, and proceeded to show the clerk three photographic lineups. The Anglo clerks picked out the correct Anglo customer from a 5-photo lineup more often than either the African-American or Mexican-American customer from those line-ups. The African American clerks were correct more often on the African American customer than either of the other two races, and finally, the Mexican-American clerks were correct on the Mexican Americans customer more often than the other races. The greater ability to recognize members of your own race compared to other races is known as the Other Race Effect. It's reliable. No one denies its existence. It is easy to replicate[4].

Why, How and Back to Why
Why does the Other Race Effect occur?

Are members of some races physically less distinct than others? No. If some groups were just plain hard to distinguish, then everyone would do poorly for those groups. The studies instead find "crossover" such that people from races that are hard to recognize by one group are easy to recognize by members of that race and vice versa.

The Other Race Effect occurs because of perceptual learning. With practice, we become better at making fine discriminations within faces. By "discrimination" I don't mean anything involving prejudice, but rather discrimination in the perceptual sense of being able to discern the differences between things that are similar. The mechanism that allows us to do this is not specific to race. You actually already saw in Doty's study that it is easier to distinguish between faces from your own nationality than it is from other nationalities. The perceptual mechanism is also not even specific to faces. For example, it is also easier to recognize a person from voice alone when the voice comes from your own familiar culture rather than from somewhere else. This is true even when it is in the same language, such as English spoken by Americans vs. the British. The generality of perceptual learning may even imply that all taxicab drivers look alike to people who do not drive taxicabs. But it is really even far more general than any of these and reflects a mechanism that occurs in absolutely all of perception.

Consider the taste of wine. If you have not had much experience tasting wine, then you have just one division, red vs. white, and maybe also good vs. bad. With practice, however, you can taste dozens and dozens of subtleties between different wines. Nearly everyone has heard of the extensive vocabulary that wine tasters have developed—terms like "dry" and "sweet", not to mention absurd-sounding labels like "angular", "toasty", and "fleshy". But it's not just that a set of pompous individuals made up lots of terms but rather that the experienced individual can now actually *taste* the differences between the wines. The labels became a way to keep track of all the different tastes that were emerging. An expert and a novice can put the same thing on their tongues, yet have different perceptual experiences—fascinating! Another example concerns cacti found in the desert. When I first moved to Arizona, it was pretty much "ouch" vs. "ok", a distinction which I suspect was even harder on my dog, Viking. Later, I was able to perform such perceptual feats as telling a teddy bear cactus from a prickly pear. A great example is reading X-rays. You may have gone to a doctor who took time enough to show you your X-ray. He points and says "see the fluid on your lungs"; you say "uh, huh", but what you are thinking is "What fluid? What *lung*? Is that a lung?" He does not have X-ray vision, so to speak, he's not a rocket scientist, and he was not born reading X-rays. He merely has seen hundreds and hundreds of X-rays and as a result, can now see things he could not before.

Think about whether you are a perceptual expert at anything. Besides expertise in faces of your own race, is there any perceptual ability that you acquired? Think especially if you have acquired a unique perceptual discrimination ability that not everyone shares. A couple of examples from people I have asked are hearing the different instruments in a concert and the differences in musical notes (which all musicians share), and feeling differences between fabrics (this unexpectedly from a young man who had a part time job in men's suits). One ability that arises frequently is hearing the differences between words in Spanish—or substitute Swedish, Russian, Mandarin, French. A foreign language sounds like fast-paced gibberish when you are not familiar with the language. Learning a second language gives you perceptual expertise that your neighbor may not have. For additional examples of perceptual expertise obtained from ordinary people, see the end of Chapter 7. Perceptual expertise is to be found in every area where there is something to perceive.

In the case of race, we become better at detecting features that are most useful for distinguishing among faces of our own race. It's not magic. We get better because of extensive experience we have in our own race. Even if you don't normally realize it, faces are very similar visually to one another—compare one to a table for instance. A face is more similar to another face than to any other object. With practice, we learn to detect what is different about the different faces that we encounter frequently. But what happens if you try to apply what you have learned to a new group? You look in all the wrong places. You don't see what you need and the features you do notice won't do you any good. Consequently, you don't detect what makes one face different from another in the new group. You are back to the default state of the novice where everyone looks alike because faces are all alike and more similar to one another than to tables.

Following are some descriptive terms that subjects in an experiment on face recognition said they used to try to recognize the faces in the experiment[5]:

Caucasian subjects: hair color, hair length, hair texture, eye color
African American subjects: hair position, eye size, eyebrows, ears

The subjects from the different races described different things. I'm not saying that these particular features are the ones that can be used to distinguish these different races or that they were even actually used by the subjects who reported them. Verbal descriptions are not always right. When asked to introspect, people usually comply. However, when people think they are doing a task one way, often what they are really doing is automatic and may be based on something else entirely. In much of perception, the learning processes involved permit only restricted access—if that—by conscious awareness. Conscious awareness may be misleading, erroneous, or even get in the way. This is something I will raise again in subsequent chapters, especially Chapter 11. But the subjects' descriptions do illustrate that even at the conscious level, different features are emphasized by different groups of people. Everyone develops and concentrates on those features that are useful for making distinctions between the faces seen every day, whether consciously or not. Your prototype of a face is based on your own race because you encountered more people from your own race. If you apply what you've learned to a different race, it is not very useful.

How does the Other Race effect occur?

At the level of the brain, several different mechanisms may be instantiating the changes that we know occur with practice. My favorite is that we "steal" neurons that would normally be doing something else. Say for the sake of argument that the distance between the eyes of a face is between 2 and 3 inches. By looking at many such faces, we can become super experts at 2 to 3 inch intervals by recruiting cells that normally would have coded distances of 6 to 7 inches; we force them to do our evil bidding and also code for 2 to 3 inches. Supersized machinery allows for supernormal or greater processing on those features that are particularly useful for our needs.

An interesting consequence of cell theft is that it predicts experts should be worse than novices at other tasks that are not in the realm of their expertise. If neurons have been completely stolen for one task then they are no longer useful for the purpose they held before. Studies of visual line detection provide evidence for this prediction. Subjects that are made to be experts through hundreds of tedious trials on distinguishing small line segments that point in oh so slightly different directions do worse at lines in novel orientations than novices to the task[6]. Doesn't this mesh well with our experiences? The math geek comes to mind; the one that is a magician with numbers but loses keys and rarely ties his shoelaces. A pathological version is the savant who has a newsworthy talent, like reproducing any song on a piano heard only once, but cannot master getting dressed, carrying on a conversation, or making change with coins. Does expertise come at a cost? There is an old phrase: "He is a jack of all trades, but master of none". More people have heard the first half of the phrase than the second half. Perhaps we have always been aware of the tradeoff between being an expert and a generalist, but modern studies suggest this is not a limitation on our time, but rather our processing capacity (and see section below on "Why again...").

When cell theft is the mechanism for acquired expertise at your own race, it may be doubly destructive for perceiving another race. Not only do the features you have become specially equipped for not do you any good for another race, but you have stolen away the equipment from those features that would! The cost of this expertise is performing worse than baseline at faces of another race[7].

In the first chapter we looked at several brain disorders affecting perception. The improvement in perception with practice here relates back to destruction of perception with damage. Stealing cells is also relevant for recovery from brain damage. If recovery occurs, the mechanism often involves the recruitment of different brain regions to take over functioning for the damaged areas. Stealing cells is involved in both the initial acquisition of expertise and the reacquisition of perceptual abilities that have been lost catastrophically. Besides the stealing of neurons, other brain mechanisms that underlie perceptual expertise include having a neuron be more selective in what it responds to (known as a sharper tuning curve) and simply an increase in the use of the relevant neurons when possible. Some of these may not have the dire implications for the cost of expertise that neuron theft does.

But I want to move on to the level of the mind—the psychological analysis. One helpful idea is that with practice, it is as if you are looking through a magnifying glass. This may only be a re-description but it nonetheless captures the experience that at first you see only the basic nature of an object of interest; later, as an expert, all sorts of details are plainly visible—as if one were looking through a magnifying glass. You realize you like-like him and suddenly you see his eyes are round and set deeply and the upper lip is curvy with an asymmetrical smile. These features had previously been invisible to you. So much so that you might swear they were never there before or wonder how you could not have seen them before.

Another useful idea is that before training, there is a many-to-one mapping between what is in the world and what you experience. At first, you cannot distinguish between all the similar things—such as faces—and you have one big wastebasket category for all of them. Perception gets more refined with experience. Different groups can now be distinguished and the categories are smaller. The categories continue to get smaller until finally individuals can be recognized. The left part of the diagram in Figure 5-9 represents when everyone looks alike. After a little practice, you may be able to determine someone is Korean rather than Chinese, but all Korean people still look alike. Later, as depicted at the right of the figure, one can actually see the differences between all the individuals. Note that I have cheated because I have not said what you have to do to become an expert. We will discuss that further after this introductory chapter (see Chapter 7). For now, you can think of this discussion as "what is

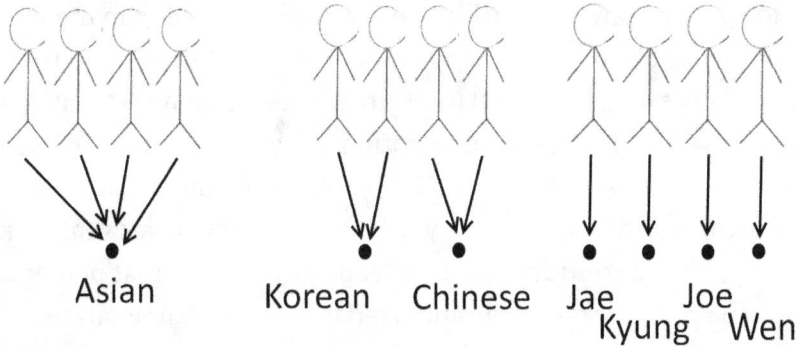

Figure 5-9 Change in the mapping between the world and perception from many-to-one to one-to-one. The perceptual categories get smaller and smaller (left to right) with experience, eventually allowing each individual to be recognized.

changing?" when you get to be a perceptual expert. Physiologically, new neurons may be recruited and psychologically, you see more than you used to. Eventually, you see all the relevant details and ideally have a one-to-one mapping between relevant items in the world and your experience of the world.

One final idea to introduce at the psychological level is the most important one. The Other Race Effect occurs because of error correction. *What's really happening in this type of perceptual learning—expertise—is that perception changes because you detect that you are failing to process things in the world.* Error correction may sound familiar because we have touched upon it in every previous chapter. Error correction is the key that unites all the different kinds of perceptual learning.

Why again does the Other Race Effect occur?

For now, I want to return to the question of "why"—this time a deeper sense of why. Why should experience cause you to perceive things you never did before? That is what happens when you become a perceptual expert on seeing faces in your own race, for example, or say, faces in Korea because you lived there for three years. Isn't perception stable, accurate, and unchanging? Shouldn't it be? A straightforward view of perception is that it should be all of these things if it is to provide a consistent reliable gateway to what is in the world. At this point you are likely more sophisticated than the typical perceiver who has not thought much about what allows them to successfully see or hear. You have already seen examples from earlier chapters where changes in perception throughout life are critical if perception is to be accurate, such as when localizing sound following head growth and feeling where your body is following movement and tool use. This is contrary to the usual prior intuition that perception should be your rock —stable and fixed forever. The unanalyzed view that many people hold is that as long as you are not blind with nothing wrong with your eyes or brain then if it is out in the world, you will see it. In case you hold any lingering intuition for these types of views, it is worth examining these naïve assumptions about perception. Begin with stability.

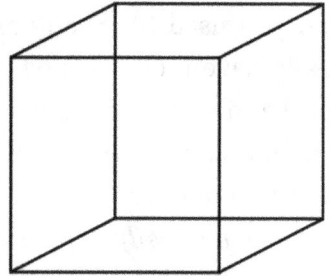

Figure 5-10 The Necker Cube was created by William Albert Necker in 1832. The cube will appear to reverse direction. Looking at certain places can encourage one perceptual interpretation or the other suggesting some role for conscious influence over perception, even if indirect.

Here is the famous Necker cube (Figure 5-10), which you may have seen before. Keep looking at the cube and try to get it to reverse direction. One instant it's facing up and to the right, and in another down and to the left. Which is it? It's your perception that is changing here, not the object in the world. In fact, this is at first so counterintuitive to how we assume we perceive that children shown the Necker cube have been known to ask "what did you do?" (anecdotally from the lab of David Premack.) It's your own perceptual system that changed what you saw though, not the world.

Now look at a painting by Dali (Figure 5-11, facing start of chapter). Looks like art... Look closer. Do you see a giant head? Keep looking. If you don't see it, look towards the left. The heads of the two normal-sized people with the big white collars are the eyes of the giant head. Their dark hair forms the eyebrows and the empty space above forms the hair and head of the giant disembodied head. It's a picture of Voltaire hidden there. Who? We really should update these. Perhaps the hidden giant head should be Justin Bieber (see cartoon in Figure 7, Chapter 7). Even when the world is held fixed, perception is not stable.

Nor is perception always accurate. Figure 5-12 starts a dull visual illusion that you can easily create before your very eyes, which makes it notable. Draw for yourself 2 vertical lines that are equal in length.

Figure 5-12 How to draw the Müller-Lyer illusion. Begin with 2 lines that you are convinced are the same height.

Now add lines pointing down or up to the ends of each line segment to look like the picture in Figure 5-13:

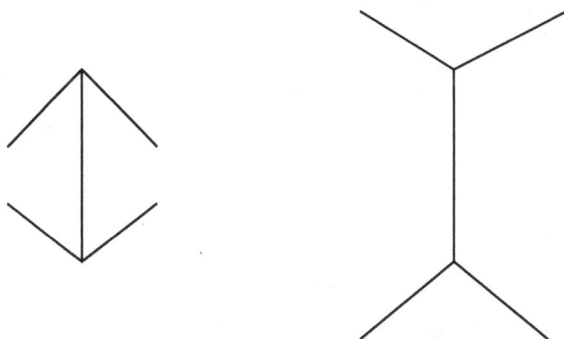

Figure 5-13 The well-known Müller-Lyer illusion was published by Franz Muller-Lyer in 1889. The two vertical lines segments are actually the same size, but we falsely perceive the line with fins extending out to be longer than the line with the fins pointing inward.

The vertical line on the right now looks like it's longer even though you *know* for a fact that it is not. You don't have to take my word for it since you created the illusion yourself. Your perceptual system is deceiving you. I have not explained why this *Müller-Lyer illusion* occurs, nor have I explained why the bistable perceptions you saw previously occur. I am just showing quickly that perception is not stable or accurate.

Finally, consider that perception is also readily pushed around by experience. For maximum effect, recruit at least one other person to join you. Think of yourself as the brain recruiting neurons away from their tasks to suit yours. You can do one half of the demonstration on your own. I am going to show you one picture for a few seconds which your friend should not look at. It is shown in Figure 5-14a. On the next page, Figure 5-14b is a picture that only your friend should look at for a few seconds. Now both of you flip to the third picture (Figure 5-14c on the page after that) at the same time and as soon as you see it, call out loudly what you see. Pause here and stop reading until you do the demo. The government is watching you, so go ahead and do it.

Figure 5-14a. Look at the picture for 5 seconds. Then bypass the next page and go Figure 5-14c on the page following. Call out what you see.

Did you call out "mouse" or "rat" while your friend called out "man"? It's especially amusing when we make 50 or 100 or even 500 people do it at the same time, with everyone sitting on the left side of the room seeing the mouse and everyone on the right seeing the man. The first picture, which is the one you saw, was a mouse. The second picture, which your friend saw, was a man. If you look at the third picture carefully, which both you and your friend saw, it's a more ambiguous picture that could be either a mouse or a man. If your own personal experience was that there was just previously a mouse out there then it's more likely to be a mouse again since there are mice in your environment. Therefore, when shown the ambiguous picture, a mouse is what you see. On the other hand, if in your own experience you just encountered a man, then the ambiguous picture is more likely to be a man so that is what you see instead. Rather than perception remaining unchanged despite our individual personal experiences, those experiences are used to keep perception as accurate as possible for your environment. If you don't find the demonstration impressive, dwell on it longer. We have already been seeing

Figure 5-14b. Look at the picture for 5 seconds. Then turn to Figure 5-14c on the next page and call out what you see.

that perception can be changed by experience. Here in just a few *seconds* of experience, I was able to push around and change what you actually *see*!

So perception is not always stable from moment to moment, nor accurate, nor unchanged by experience. We can consider now that the Other Race Effect not only results from experience but that it is the norm for perception to be changed by experience: it is adaptive. There is too much information in the world to process and therefore you must use only what you need. The general learning mechanism that gives rise to the Other Race Effect allows you to focus on the perceptual information that is most relevant in your environment. The world is filled with endless amounts of information. Look around the room and try to describe everything that you see. It's a long list and if you think you are done, focus carefully on one thing and note how many more things there are to notice about it. It's a never-ending process.

Figure 5-14c Call out what you see. Whether you see a mouse or a man depends on which picture you looked at before this one.
Figures 5-14 a–c from Gleitman, H. , *Psychology* , Norton who adapted it from B. R. Bugelski, B.R & and D. A. Alampay, D.A (1961), The Role of Frequency in Developing Perceptual Sets, *Canadian Journal of Psychology*, 15, 205-211. Canadian Psychological Association. Reprinted with permission. *

Not convinced that you don't see everything there is to see? All Americans have grown up with the (so called) monetary unit of the penny. Can you form an image in your mind of what the heads side of a penny looks like? When you have that image in your mind's eye, answer the following questions: 1) Is Lincoln facing left or right? 2) Is there anything written above Lincoln's head and if so, what? 3) Is there anything written below him and if so, what? 4) Is there anything to the left? What? 5) And finally, is there anything to the right? What? Did you find these questions easy or hard to answer? Look at a picture of a penny in Figure 5-15 on the next page to see how well you did, (Did you instinctively want to go and get a penny yourself rather than trust my picture? If yes, then congratulations; you are thinking for yourself rather than taking my word for it. You never know when I could make a mistake —or intentionally trick you.) No ordinary person gets all 5 questions right. You have seen pennies most days of your life and you think you know what they look like, yet you don't know many of its features and don't know that you don't know. That's because there is just too much information out there. It's true for any object or collection of objects in a scene that you encounter. And that was just one single penny that has too many visual details to process. One human face contains even more information. There are hundreds of features and relations in a face. To name just a handful: the distance between the eyes, distances from nose to top lip, distance of eye to left corner of lip, and the distance from eyebrow to chin. Wax museums use over 100 measurements just to recreate the face. There is too much information in the world to perceive all of it and you need experience to tell you what to see and what to ignore.

It is important to realize that the failure to process things is *not* just a processing limitation (despite the neuron theft issue discussed earlier). *You would not want to perceive all those differences even if you could.* This is actually a very deep point, perhaps the deepest of this introduction to the Other Race Effect. Also deep was the point about error correction. Consider Tony, a student from my class. I'm

Figure 5-15 How well do you know a penny?

picking him because one day Tony came to class with every strand of his full long black head of hair completely gone. Who is this kid who is sitting in my (atypically) small class that I have never seen before? For some reason, I decided his name was Joe. I did figure out eventually it was Tony, but even after I got to know Tony quite well through two classes, research in my lab, a job in my home, a vertigo attack (mine) and life-threatening allergy attack (his), he always seemed to be two people, Tony and Joe, and it was hard to fuse them back into one. But it could be anyone and the point is this. Tony, or you, or anyone, or anything looks different when I am standing right in front of him then he does when I am on the left side at an angle. In addition, the visual stimulus is different when I'm on the right side of him, or when he looks like he's about to fall asleep in class, or when he stands up and is above me, or when he smiles. What if I interpreted all those different images as different: Front view Tony, side view Tony, sleeping Tony, smiling Tony and so on. With the exception of extreme cases like sudden baldness, we do not use the visual differences that are always there to infer different objects. To do so would be catastrophic. Can you imagine? Even though the visual images of each view are different, we don't want a left Tony and a right Tony, we just want Tony. *The easiest way to achieve this is not to see the visual differences that exist at all.* Experience tells us which features need to be seen and distinguished from one another and just as critically, which don't. No two stimuli in the world are ever exactly the same, strictly speaking.

A lamp you see this second is not even identical to the one you see 5 seconds from now. It is not a matter of there being a right answer to the question of "are x and y the same or different" or "are these 2 photos the same or different?" The answer depends on what you need the information for and that will change depending on the circumstances. Therefore, differences that you see or do not see between people have to be plastic and that's why it's governed by experience. For pictures of Tony with and without hair, see Chapter 12.

Implications for Eyewitness Testimony

It may make you uncomfortable to know that you are not so good at distinguishing between people from a racial group different than your own. But the implications are far greater than social awkwardness. Consider what happens when witnessing a crime that involves a person from a different race. If only it was as simple as saying, "I'm not sure" when shown a line-up of suspects, there would not be nearly as big of a problem. However, it is very easy to be 100% certain of a recognition, yet misidentify the wrong person as having committed the crime. Juries are swayed by confidence and such testimony can be compelling. Someone who boldly says "I saw it with my own eyes" and "I'm positive that was the guy with the gun" makes both for successful television drama and courtroom conviction. However, you can be certain—yet wrong. If you can't see any difference between two things then you are going to be confident that they are the same even though they are not. This leads to a greater chance of misidentifying someone if they are from a different race than your own because you cannot see the difference between people.

Many convictions that are being overturned because of DNA evidence turn out to have been based on cross-racial identification. In Arizona, Larry Youngblood was an African-American man, then 30 years old, accused of child molestation, sexual assault, and kidnapping. A 10-year-old boy, David L., had been snatched from a carnival and repeatedly sodomized over an hour and a half period. The boy was molested first in a secluded area, then taken in the assailant's car to the assailant's home, tied up when the car wouldn't start, and molested again before being taken back to the carnival. David was Caucasian, and described his molester as a black man with a damaged eye. One officer remembered that he had contact with a black man with a defective eye, Larry Youngblood. David was shown a photographic lineup of Youngblood and 5 other black men, all 6 of whom had one eye blocked out with ink on the photos. He picked Youngblood's picture as the man who molested him, said he was "pretty sure", put on his glasses, and said he was "real sure". During the trial, David was certain that Youngblood was the man that assaulted him and his testimony was the primary evidence in the case. Youngblood was found guilty. In August 2000, improved DNA analysis was able to rule out Youngblood as the source of the semen that was part of the original evidence. The falsely accused man was exonerated of the crime—but it was too late, as he had already served his entire 10 years. The DNA from the semen was later matched from a national database to Walter Cruise, a black man who also had a defective eye. Remember that as noted in a previous chapter: Perception is lazy. The perceptual system, like the jury that was to come later, thought that it's burden of proof for identification was filled by a "black man with a bad eye" and did not need to look for any further information.[9]

I have testified as an expert witness in court cases about the Other Race Effect. In Arizona v. Andre Smith, I argued that two Caucasian witnesses could easily have been mistaken, despite their best intentions, that the black man they saw firing the gun outside the nightclub was Smith, The witnesses had no particular expertise with the African American population without relatives or close friends of that race, making them vulnerable to the Other Race Effect. In addition, they would have been focused on the weapon during the crime, which research has shown makes accurate identification even harder. Finally, the identification for both witnesses was based on a *show-up* identification in which the police brought by one suspect in a police car asking, "Is this him?" It is even easier to mistake one hard-to-recognize man for another if you trust the police have brought you the right person and you don't even have to pick him out from similar looking photos. The eyewitness accounts were the only evidence against Smith and he was found not guilty of a crime that could have led to a 35-year prison sentence.

It's important to educate juries that it is not racist to do poorly at recognizing people from other races, but rather results from a general perceptual phenomenon based on experience. It is also informative to explain that in no way does this mean the witness is lying, but rather that people can easily be confident and not know what it is that they don't know. Should you ever find yourself testifying as an expert witness, some tips given to me by defense attorney Anthony Payson are to look at the jury while speaking (to make a connection), to pause when the prosecutor asks a question (in case the defense objects), and not to touch your chin (because they are taught this gesture reflects deception !)[10]

How big an effect is cross-racial misidentification? Ask yourself first if you are so good at your own race. Look again at Figure 5-8. If your own race is shown, is it easy for you to tell the difference between the individuals when you have just the face and not the easily changeable superficial features like hair and clothing? There is a great deal of individual difference in the ability to recognize faces.

Or consider the upside picture in Figure 5-16 (facing the start of the chapter). It might look maybe a little strange to you if you notice anything unusual at all. But to see just how strange, you will have to turn the book upside down. Alda's true face is adjacent for comparison. Righting the upside down face reveals that it is almost inhuman—positively monstrous (sorry Alan!)—yet you did not see this when it was inverted. The mouth and the eyes were pasted in upside down relative to the rest of the face (that is, they are objectively right side up...) but when the whole face is upside down, you do not notice the misoriented features or just what a big impact they have. This is known as the *Thatcher illusion.*[11] Face perception is fascinating and complex. The literature is full of intriguing findings. It is also cluttered and a mess. Add to that problems with eyewitness accounts and the bottom line is that an eyewitness from the same race cannot always be trusted either. Cross-racial identification clearly adds further problems to eyewitness reliability but putting a consistent number on it has been difficult.

Another issue we do not know much about is how much experience is needed to become an expert on Pakastani faces, or Native American faces, or Russian faces, and so on. Currently we cannot really tell if a witness will be good or bad at other races if they have had some contact with that race, but no intimate relationships or childhood friends. Can we train people to become better witnesses? Since the Other Race Effect results from differential experience with different race faces, then yes, it should be possible to do so. We have been doing this in our own lab with principles taken from the general non-face perceptual expertise literature. For instance, having subjects attend to one specific feature when they are learning faces should work and indeed we are finding it does. I will return to this research in Chapter 7. You would think there would be tons of work on this topic, but surprisingly only recently has the Other Race Effect been examined rigorously from a perceptual learning view. Training large groups of people is currently not practical and won't help with ongoing cases of eyewitness accounts. Eventually, it would be great to see such training in all middle schools the way second languages are taught. For now, the best we can do is to inform as many juries, judges, and lawyers as possible about what the data show. It is also not unreasonable to fashion a face test to give to witnesses to assess their abilities. Perhaps as more convictions are overturned with a witness misidentification component, the legal system will be more receptive to this idea.

Conclusion

It is not a myth that it is easier to identify people of your own race compared to identifying people in other races. In addition, it reflects a general fact about how perception gets more precise with experience that applies to wine, X-rays, and anything you can think of, and not just faces or races. It is adaptive to see more with experience because there is too much information in the world to want to detect it all. Finally, be especially doubtful of cross-racial identifications in the context of eyewitness accounts of crimes.

This information in this chapter is usually the introduction I give to professionals involved in law enforcement minus a couple of the more abstract points. These groups not uncommonly know little about perception. In these pages, we have an opportunity to consider aspects of the Other Race Effect in more depth. Consequently:

Questions

1.   How could you prove that the Other Race Effect is not because some races are more easily distinguishable than others?

2.   How could you prove that the Other Race Effect is not "genetic"? That is, how would you show it's not that you are born being able to distinguish your own genetic race and not others?

3.   Suppose you have pictures of 20 Caucasian-American faces and 20 Latino faces. You want to study the Other Race Effect, but you are concerned that by chance, the 20 Latino pictures happen to be easier. What might you do to solve this problem?

The Other Race Effect and Perceptual Learning, Gibson style

Figure 6-1 Do animals have trouble distinguishing between "people" from another species? Original cartoon by Felice Bedford and Whitney McNeil.

E njoy the cartoon in Figure 6-1 while you can because there is much work to be done. The first half of the chapter delves into details of the Other Race Effect, a specific example of perceptual expertise involving faces. The second half of the chapter transitions to expertise throughout the rest of perception. Not detail oriented? The next chapter is also about perceptual expertise, but brings us back down to earth.

PART I THE OTHER RACE EFFECT

Answers to Questions on the Other Race Effect

1. *Do we have a harder time perceiving people from some races only because the people in those races are intrinsically more similar looking to one another? How could you prove otherwise?*

If some races were just harder to distinguish than others then everyone would do worse on those races—even people from that race. Suppose people from Pakistan find Korean faces harder to recognize than Pakistani faces. If this occurred because Koreans were harder to tell apart than Pakistanis then everyone would do worse on Korean faces, including Koreans themselves. However, in the real world, Koreans do no such thing. They find other Korean people easiest to recognize. To prove that a purported Other Race Effect did not occur just because some races are harder to distinguish than others, one needs to find this "cross-over" or reversal such that what's harder for one group is easier for another and vice versa. Getting some kind of crossover is critical, although it need not be a perfectly symmetrical reversal. In abstract notation, if race A finds B harder than A, then to prove that is not simply because B *is* harder than A, show that when race B is doing the observing, they find B to be easier than A, not harder.

2. *Are we better at perceiving our own race because we are genetically pre-wired to do so, perhaps with perceptual templates of our own race? How could you prove otherwise?*

To separate genes from environment, try exploring face recognition following adoption. Assess how an adopted child or adult does at faces of his adopted race compared to someone who is not adopted:

	Born	Raised	Test
Group 1 (adopted)	Race A	Race B	Race B
Group 2 (not adopted)	Race A	Race A	Race B

If the adoptee does better, we know that experience led to facial expertise. However, we have not fully answered the question. Experience could help with other races, but we could still also be born with a genetic ability to recognize our own race. Consider then also testing the natively born race:

	Born	Raised	Test
Group 1 (adopted)	Race A	Race B	Race A
Group 2 (not adopted)	Race A	Race A	Race A

The universe of possible outcomes to this test are that the adopted person (Group 1) could do better, the same, or worse on recognizing his own native faces than the non-adopted person (Group 2). There is no model that predicts he would do better and we need not consider that possibility further. If he performs the same then he is doing just as well on his native race even without seeing them in his environment. This could, if it occurred, argue for a genetic advantage for recognizing faces in the native race. Yet such a conclusion would not be entirely compelling because the adoptee could get experience just by looking in a mirror. It is not yet known whether looking at just one face (you own!) every single day is sufficient to produce expertise at your race. Further experimentation would be needed to tease apart genetics from environment if this outcome occurred. If he instead does worse at his own race than a non-adopted person, this would show that experience is necessary to be as good at recognition within one's own race as someone who gets that experience. Such a result would provide evidence against a hypothesis that the Other Race Effect is genetic. The two experiments I described would need to be done with groups consisting of many people but I have described them with a single individual for ease of exposition.

There are many other permutations that can be tried and each would reveal another piece of the puzzle. One comparison that is tempting to make but would *not* allow a valid comparison would be:

	Born	Raised	Test
Group 1 (adopted)	Race A	Race B	Race A vs. Race B

It seems almost irresistible to ask how an adopted person does at faces of his own race that he did not grow up with vs. faces from the different race that he did grow up. Unfortunately, it would be impossible to know how to interpret any result that happened here. If he does better at one race than another—regardless of which race that is—couldn't the reason just be that the pictures chosen for testing in the laboratory happened to be easier? There is no other group to compare the data to in order to rule out this artifact. To follow through with this approach, many other groups would be needed to make higher order comparisons.

Incidentally, to my knowledge, no one has ever found any evidence for a genetic contribution to the Other Race Effect. Think of language acquisition as a parallel situation. You learn whatever language you were exposed to, even if it's not the language of your ancestors.

If one set of pictures is indeed harder than another then that leads to the third and final question:

3. *What you would do if it turns out the set of pictures you were using for one race turned out to be harder than the pictures for another race?*

Suppose the yearbook you used for Caucasian pictures were somewhat blurry. Or the pictures you had for Latino individuals happened by chance to contain some very unique looking people. I ask because this happened to us and I'm curious what you would try to do about it. Our pictures weren't blurry or especially unique, but for whatever reason, one new set of pictures we gathered were harder pictures for everyone (of all races) to recognize and harder than previous sets we had used. How should this practical obstacle be overcome? We decided that there are two basic approaches that can be taken. I ask students all three of these questions, including this last one. As I mentioned before, I especially like asking questions where others with a fresh perspective might know something that we do not. All the answers I received on this picture question fall into only one of the two possible approaches. One group of students suggested increasing the number of pictures from 20 to 50 for each race. Other students suggested removing the hair for the people in the Latino pictures. Both of these suggestions work on equating the difficulty of the two sets of pictures. Increasing the number of pictures would decrease the odds of picking by chance predominately unusual individuals and removing hair would make the easier pictures harder. Equalizing the difficulty of the pictures is one of the solutions to mismatched pictures. Researchers usually make some attempts like these to get picture sets as close as possible to eliminate any bias.

The second approach is actually to live with it! This is because it may not ever be possible to completely equate two relatively small sets of pictures anyway. The Other Race Effect can still be demonstrated. Consider these hypothetical data (Figure 6-2):

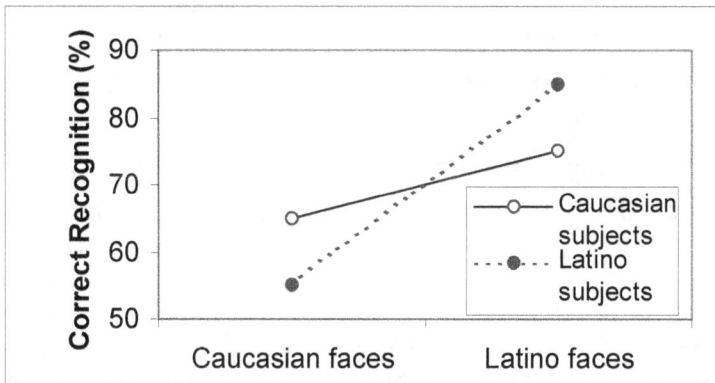

Figure 6-2 Hypothetical data depicting cross-race identification between Latino and Caucasian (Anglo) people. Note how everyone is doing better on recognizing the pictures of Latino faces, yet there is still a "cross-over" between the races.

Inspect the graph to see that everyone is doing better at the Latino pictures. Look first at Caucasian subjects, shown by the solid line. They do better at recognizing the Latino pictures. (Note how the right side of the line is higher than the left side and higher means greater recognition.) Now look at the Latino subjects, shown by the dotted line. They, too, do better at recognizing the Latino pictures. (Note the right side of that line is also higher than the left side.) But these data would still show the Other Race Effect. There is still a crossover of the two races. For the Caucasian *pictures*, Caucasian subjects are doing better than the Latino subjects. (Compare now the open and closed circles on the left side of the graph.) And for the Latino pictures, Latino subjects are still doing better than Caucasian subjects. (Compare the open and closed circles on the right side of the graph.) So even though everyone does better on the Latino pictures, each race is doing better than the other race on their own race pictures. Better performance for everyone on the Latino pictures would likely implicate that the Latino picture set happened by chance to be easier. As just seen, this does not prevent the ORE from being detectable. In fact, even just one half of a crossover suggests an Other Race Effect, but I won't go into the details.

Everyone should have gotten the essence of question 1 correct by looking at the previous chapter. For question 2, anyone who had the sense that the environment must be separated from genes in some way was essentially correct. Specific experiments always require some fussing to get them just right. If you like comparing yourself to norms, typically about half the students get at least two out of the three questions essentially correct.

An Experiment on Sex and the Other Race Effect

The experiment on sex is not nearly as interesting as you are hoping, but still interesting. Look at the photo which is facing the start of this chapter (Figure 6-3). The photo shows a Bushmen elder from Nambia. Bushmen are known for the unique click sounds in their languages and are the oldest inhabitants of the planet. We are all genetically descendent from them. Can you tell *without hesitation* if the photo is of a man or a woman? (Answer is at the end of the book in the Notes section[1].) Alice O'Toole and colleagues wanted to know if the Other Race Effect extends to perceiving the sex of a person[2].

Since it's harder to recognize a person from a different race than your own, they asked whether it would also be harder to see if someone were male or female if they were from a different race than your own. Here is what they say in the introduction of the article. But don't assume that everything they say is correct.

Specifically, to recognize faces we must be able to encode something unique about them. By contrast, to discriminate faces along sex, race, or age dimensions we must be able to perceive the features in individual faces that are common to an entire group of faces, but which also serve to distinguish a face group from alternative groups(s) (e.g. features that all male faces have, but which female faces typically do not have). In other words, these tasks involve a perceptual discrimination between categories of faces.

O'Toole and colleagues are arguing that recognizing an individual on the one hand and determining an individual's sex on the other are fundamentally different because recognition requires that unique features be processed whereas determining the sex requires distinguishing between categories. This distinction is not quite right. You are actually in a position to generate a very sophisticated argument. I am doubtful that all of my colleagues could do so. Remember the discussion on the many Tonys and how each looks different? In order to perceive just one Tony we have to ignore the differences between the different Tonys and categorize. We take all the features that are common to Tony and serve to distinguish him from an alternative set of features—that comprise, say, Paul—and put those in one category. This is exactly what the authors describe for distinguishing male from female faces. In contrast to what the authors' claim, both determining the sex of a person *and* recognition of a person require categorization. Both require that some features be used and some be ignored. We ignore the differences between Tony and Paul to categorize them both as male and we ignore the differences between Left Tony (Tony seen from the left) and Right Tony to categorize them both as Tony. Does it matter for the experiment that face recognition and gender recognition do not differ in this way? No, but that is no excuse for ever accepting a bad argument. You never know what might end up being crucial at a later time.

What the researchers did for the experiment itself was to show 60 Caucasian pictures and 60 Japanese pictures to Caucasian and Asian subjects. Half of the pictures were of women and half were of men. They removed the ability of hairstyle to reveal the gender by embedding each face in an oval shape, which cropped off much of the hair. The Asian subjects included people who were Korean, Chinese, or Vietnamese, though not Japanese. Each face was shown for 75 milliseconds—which is less than $1/10^{th}$ of a second—and subjects were required to indicate whether the individual was male or female. They found that the Other Race Effect does indeed extend to making judgments about the sex of the individual. Subjects were better at determining sex for own-race faces than for other-race faces.

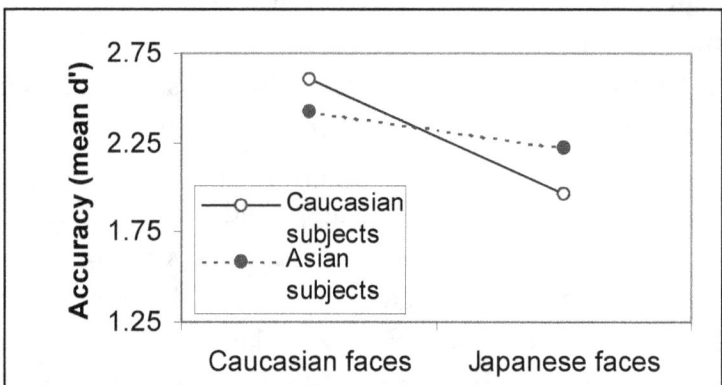

Figure 6-4 The Other Race Effect extends to sex judgments. Subjects in O'Toole, Peterson, & Deffenbacher's experiment indicted whether each picture was of a man or a woman.

I have taken the data they collected and graphed it in Figure 6-4. You may note that the Y-axis lists mean d' as the measure of how well subjects did. Unfortunately d' (pronounced d "prime") is not a very intuitive measure for most people, including my colleagues, so when we see that Caucasian subjects scored a d' of 2.60 on the Caucasian faces there is no intuition about how good or badly they did. It is legitimate to convert the data to d', but it would have been better to first see the more intuitive raw means, like percent correct, before converting the data.

Converting results to d' is based on Signal Detection Theory and is designed to get around the problem that different subjects may have different biases about how they respond. Suppose Tweedledee always says the picture is female for all 120 pictures. Suppose Tweedledum always says the picture is male for all 120 pictures. Each subject will be right half the time—at chance—since half the pictures are male and half female. But they are right differently, each with a different bias. There are four different possible outcomes: a picture can be male and the subject correctly says male, a picture can be male but the subject incorrectly says female, a picture can be female and the subject correctly says female and finally, a picture can be female but the subject incorrectly says male. Signal detection theory allows biases a person has to say male or female to be separated from the true ability to detect gender. The measure d' reflects that true ability, without the bias. So it is a legitimate, good, respected conversion of data, but it has the disadvantage of making the data not very transparent to the burning question we have of how often subjects were correct in the experiment. The researchers show a graph of their data, but it's converted even further than the one I presented in Figure 6-4.

Another quibble: I wish they had they also tested the standard Other Race Effect in this same experiment using these pictures and subjects. Then we could compare the sex judgment data to person recognition data side by side and know better how to think about these sex judgment results. Was the cross-race sex effect small compared to cross-race person identification? Or maybe it's vastly greater. So there's a lot we don't know about the results of this study, but all experiments have advantages and disadvantages and we can learn from the advantages.

If you inspect Figure 6-4, you will see that everyone did worse on the Japanese faces than on the Caucasian faces, even the Asian subjects. But there is still a crossover. Look at the faces rather than the races to see that the sex of Caucasian faces were determined more accurately by the Caucasian subjects while the sex of the Japanese faces were determined more accurately by the Asian subjects. The outcome is analogous to the situation discussed earlier with Figure 6-2 where, hypothetically, Latino pictures were all easier[3] but there was still an own-race advantage. Another fact O'Toole and colleagues uncovered is that women did better than men. The groups in order of accuracy from best to worst were: female Asian subjects judging Japanese pictures, female Caucasian subjects judging Caucasian pictures, female Caucasian judging Japanese pictures, female Asian judging Caucasian pictures, male Caucasians judging Caucasian pictures, male Asians judging Japanese pictures, male Asian judging Caucasian pictures, and finally male Caucasian judging Japanese pictures. This does not mean that women are better at perceiving other women than they are at perceiving men (although to complicate matters, some researchers do find this), but rather that they do better on all the pictures than men do. An advantage for women when it comes to perceiving faces generally is found in some studies but not others.

If the article were open for discussion, is there anything else that you would want to question? One issue that tends to come up is the 75 msec presentation of the pictures. A number of students have been concerned that those times are so fast that the study has nothing to do with real life. We see people for a lot longer than the tenth of a second that can be created artificially in a laboratory experiment. I always find this comment interesting because this aspect of the study did not bother me. As researchers, we are used to pushing the limits of the system being studied. Had they used naturally long times for showing the faces instead of the unnaturally restricted times then everyone would have gotten the gender correct 100% of the time and nothing would be learned. Making it taxing for perceptual systems revealed the inner limitations that otherwise would not have been visible. But do the short presentation times nonetheless render the findings useless for the real world? I hope it still applies to the real world, but it's hard to know. Another issue that comes up is that Japanese faces were shown in the study but the Asian subjects were not Japanese. This mismatch was not present for the Caucasian side of the experiment because both the faces and the subjects were Caucasian. This in turn leads to discussion of exactly what one means by Caucasian and Asian anyway. Issues of race are always complex, but the mismatch at least was not a problem for this study. Had the non-Japanese Asian subjects been found *not* to be better than Caucasian subjects at the Japanese faces, then the mismatch could have been the reason. But since they found that cross-racial effect, it turned out not to be a problem.

Overall, O'Toole and colleagues' experiment shows that the Other Race Effect between Caucasian and Asians extends to judging the sex of an individual. The original Other Race Effect refers to being able to recognize an individual more accurately when they are from your own race compared to a different race. This addendum shows that we can also more readily determine if a person is male or female when they are from our own race.

Alice O'Toole has generally done a great deal for the study of the Other Race Effect. She explicitly mentions that it is an example of perceptual discrimination learning (perceptual expertise). Although it has been claimed in print for 100 years that experience causes the advantage for perceiving one's own race, almost no one explicitly ties the phenomenon to this well-known type of perceptual learning. As discussed in the previous chapter, perceiving own-race faces, wine, cacti, and X-rays are all examples of the same general learning mechanism throughout perception that enables expertise. By being explicit about the connection, principles that have been learned in the perceptual expertise field should now be applicable to the Other Race Effect. This is something our own lab is doing experimentally. But let's go to the very first original source of the perceptual expertise field.

PART II TRANSITIONING TO EXPERTISE IN ALL OF PERCEPTION

Pearls of Wisdom from the Classic 1955 Article by Jackie and James Gibson

The quick Cliff-notes history is that James Gibson had already written a radical book in perception that contributed to his then celebrity perception status. Jackie (Eleanor) Gibson was interested in perceptual development and was soon to be the co-author of an oft-cited delightful study with babies and baby animals on what everyone knows of as the "visual cliff". They joined forces academically (having already done so matrimonially) and produced two children and an influential article, which Jackie later expanded on (the article not the children) in the book *Principles of Perceptual Learning and Development*. Her work gave rise to the area of perceptual expertise and she is still cited frequently.

In the original article, the Gibsons introduce the example of wine tasting, which has come to exemplify perceptual expertise. I succumb to that, too; in the introduction to the Other Race Effect in Chapter 5, I mention wine tasting as another example of the underlying perceptual learning process. The Gibsons say further that given the exact same range of stimulation by the world, there can be two very different kinds of reaction. They describe the "crude fellow" who can identify sherry, champagne, white wine, and red wine. He has four percepts. Compare that to the "discriminating gentleman", they ask us to do, who is able to identify a dozen different types of sherries, each with numerous varieties and blends. He has four thousand percepts. Substitute beer for sherry and you may have some sense of what it feels like to be a perceptual expert. "Perceptual expertise" is modern terminology. The Gibsons describe the second individual as one who "discriminates more of the variables" following extensive experience. Here is a list of all the ways that they say someone is different before and after perceptual discrimination learning that we can extract from their paper. If you get bored, think of something you are perceptual expert at and see if you can apply any of these items. You may not find that all of them resonate well or are even useful, but it's important to keep track of everything they refer to. This way, if you later come across modern studies, you can determine what subsequent researchers actually got from the Gibsons' work, what they filled in themselves, added, or sometimes even falsely attributed to the Gibsons. People have a way of citing without actually reading the founding work in a field. It's not malicious, but one researcher may cite another from two years ago, assuming they are correct, who in turn was citing an article from three years prior and so on. After fifty years, the results are one long game of telephone where the message may get distorted.

As a result of perceptual expertise, an individual, such as the dis-criminating gentleman who knows his sherry, will, will be, or will have the following:

- respond to variables of stimulation not previously responded to
- improve because learning is always supposed to be a matter of improvement **
- in closer touch with the environment
- an increase in the number of distinct percepts
- an increase in the number of specific objects to which those percepts refer to ("to which they are specific")
- differentiate previously vague impressions ("is a matter of dif-ferentiation")
- a greater number of differential responses
- a progressively greater correspondence with stimulation **

One other comment about exact terminology concerns terms they use for the whole area—not perceptual expertise, but "specificity theory", "differentiation", and "perceptual learning". *Specificity theory* is not currently used. The Gibsons had used the term to distinguish it from other theories of perception and learning that were prevalent at the time (which they called *enrichment theories).* The spirit of the times had been that the information in the world that was available to eyes, ears and the rest of the senses were deemed insufficient to produce a unique correct percept of the world. Perception, therefore, involved adding something of one's own to the impoverished stimuli. The Gibsons argued that nothing needed to be enriched or added (enrichment theory) but instead stimulation was already too rich and needed to be made more specific or narrowed down (thus, specificity theory). In the learning areas outside of perception, there was a similar trend. All of learning was viewed as forming associations—adding new connections. The Gibsons argued for this new kind of non-associative learning process. It is hard to put you in a time machine (putting concepts to be discussed in Chapter 10 aside) so you could better appreciate how radical a view they were presenting for the 1950s. It turns out both views of perception and learning are correct—stimulation is both impoverished and too rich—but it would take us too far afield to discuss them. It would also take us to far afield to discuss that they were anti-Nativist and were presenting their views as a theory of how perception develops from experience in the first place rather than being innately given. Their agenda was different than ours but we can stick to our goal of understanding all the ways that experience affects perception and still learn a great deal from their pioneering work.

As for their other terms, *differentiation* is used sometimes, but is hard to say, as you may be able to verify. That left *perceptual learning*, which is why "perceptual learning" is used often for this Gibsonian type of learning. Unfortunately, it's the wrong term. This was noted briefly in Chapter 1 and there will be more on this point later. I have been using the term "perceptual expertise" (with an occasional perceptual discrimination learning or Gibsonian perceptual learning) to refer to this process and will continue to do so.

I want to inspect each item on the extracted list briefly and try to apply it to the Other Race Effect. Consider Erin who has gotten extensive experience viewing faces of the Tohono O'odham Native

American nation in Arizona because she does research on astigmatism in that group. For the first item in the list, perhaps Erin notices facial features not seen before. For the second item, as a first pass, we can say that Erin is improved at seeing faces. The second item is actually very important, so I put stars next to it. You may remember in the very first chapter I mentioned how improvement is critical for learning of any kind. I will come back to this, too, when discussing general theories in Chapter 12. For the third item, Erin is able to recognize more of the individual people that are in the Tohono O'odham nation, which puts her in closer touch with that world. For the fourth item, since she can now recognize more individuals, she is able to have percepts of more faces than ever before. Erin is like the wine taster who has four thousand percepts.

To continue down the list, fifth, each face percept she does have corresponds to one person out in the world. Contrast that with what happened before she had experience when each percept corresponded to many people because she could not tell them apart. Sixth, she can now differentiate between faces that previously looked identical or very similar. Seventh, distinguishing more faces and having more face percepts should enable a greater number of distinct responses, such as verbal labels. In the same way that tasting different wines can lead to an extensive vocabulary for describing each sensation, expertise at faces should allow this as well. Erin may be able to develop and use phrases such as "wise eyes" and "strong-chinned". The eighth and final item on the list also gets starred. This seems an especially useful comment because a greater correspondence basically describes progressing from a many-to-one mapping between the world and perception to a one-to-one mapping between the world and perception. This idea was helpful when introducing the Other Race Effect in the previous chapter. Initially, lots of faces in the world are missed by lumping them together in one big perceptual category, such as "Native American". With experience, perception has a greater match with what is in the world and the number of perceptual categories increases.

You may be thinking that there is a lot of redundancy in these eight items and that they may not be all that helpful for describing what perception is like for a post-experience expert. If you are still bored after thinking about how well they would do on a perceptual expertise of your own, consider it as earning points towards your history merit badge. Remember, they were developing a theory—a whole field actually—and presumably struggling how best to capture what was happening to perception. I think they do quite well by the discussion and say it best there: *"A stimulus starts out by being indistinguishable from a whole class of items in the stimulus universe tested, and ends up by being distinguishable from all of these."*

Try Their Experiment (Your History Merit Badge)

I don't know if you would consider this a reward or not, but I have created a version of the Gibson and Gibson experiment that you can try. Below is an item (Figure 6-5) for you to study, but read these instructions first. View the item for 5 seconds. On subsequent pages, there will be more stimuli, between 1 and 9 per page, unlabelled and scattered about in different positions. For each one, ask yourself: "Is this the exact same stimulus as the one I tried to remember?" Write down "yes" or "no". Repeat that for every stimulus shown by quickly going through each one until there aren't any more little pictures. When you get to the end, you will have completed one run, or "trial" to use Gibson and Gibson's terminology. (Current usage of the term "trial" differs a bit.) You will need a total of 4 runs/trials. The remaining 3 are on a website[5], though if you cannot get there, you can repeat this set 3 more times.

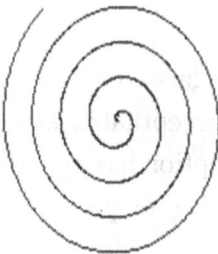

Figure 6-5 Training stimulus for an experiment modeled after Gibson and Gibson's landmark 1955 experiment on perceptual discrimination learning. Look at the stimulus for 5 seconds then turn to the next pages. Once you turn the page, do not turn back until you have judged all the little test pictures.

After I give you the right answers, you will want to create graphs

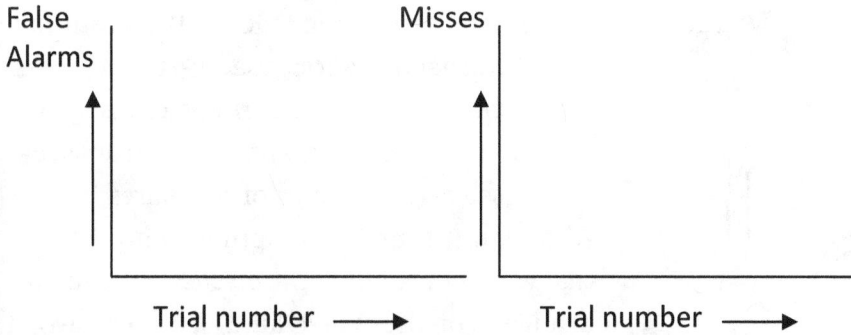

Figure 6-6 Graph format for plotting errors. Use these as a general template to graph your own performance on the Gibson and Gibson experiment replication.

of how you did. Figure 6-6 contains blank templates you can use. Each graph will have 4 data points, 1 for each run/trial. On the left graph, you will be plotting the number of errors where you thought it was the training stimulus but it actually was not (False alarms). You will be plotting your false alarms (Y-axis) as a function of trial number (X-axis). That is, you will be scoring the number of false alarm errors you made for each of the 4 runs/trials you did and putting those 4 data points in the graph. For the second graph, you will plot errors that you made where you thought a stimulus was not the same as the training stimulus, but actually it was (Misses). If you have someone else to test, too, by all means do so and plot the average of both of your scores instead of your individual scores.

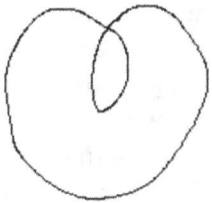

Analyzing How You Did in the Experiment

To score your data, you need to know the correct answers. Within these pages, there are 4 repetitions of the exact same stimulus as the training stimulus (you should have said "yes" to these) and 29 stimuli that tried to trick you (correct answer: "no"). The positions of the identical ones are listed in Notes at the end of the book[5]. Use that to assess if you got the right or wrong answer for each of the 33 stimuli.

When you get one wrong, be sure to keep track of whether it was a false alarm or a miss so that you can tabulate and graph those errors separately (see also the types of responses below). Although it must be tempting to not bother, that would make all of the discussion seem like drudgery of the worst kind. You also won't learn anything. In contrast, making the effort will also lead to the discovery that playing with data is quite rewarding.

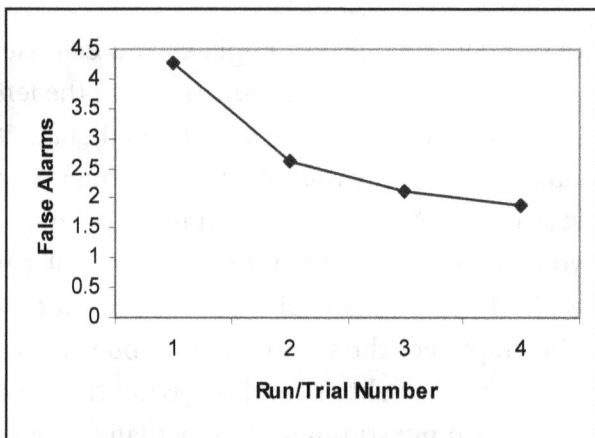

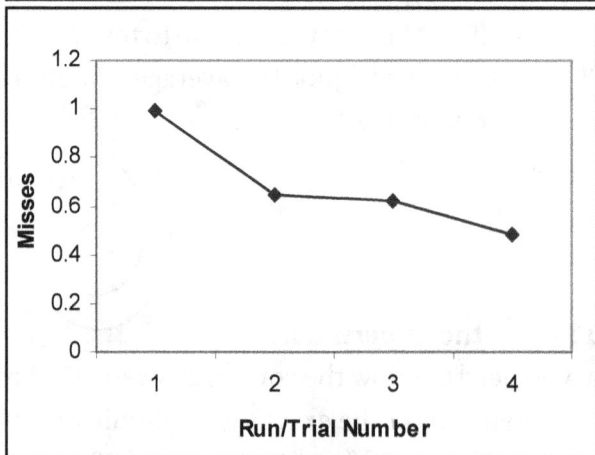

Figure 6-7 Errors in identifying the target in our replication of Gibsons' 1955 spiral discrimination task. We showed stimuli one at a time on a computer with white space in the background to thirty-four subjects. The spiral used for training was shown for 5 sec and then each test stimulus shown for 1 sec, with a blank white screen (1 sec) in between. Each of the 4 Runs began with a 5 sec view of the training image. Practice with an unrelated heart shape was given before any spirals.

Unfortunately, I can't analyze your data, but I can still show you data I collected with these stimuli (Figure 6-7) that you can compare your data to.

If you had any trouble creating graphs from your own data, consider the four types of possible responses:

Target present?	Response	Response type
Yes	Yes	Hit
Yes	Yes	Miss
No	Yes	False Alarm
No	No	Correct Rejection

For each of the 33 individual trials in the first whole run/trial, indicate which *Response type* you had. Then add up all the ones that were false alarms and that number is the first data point of the first graph. Then consider your second run/trial through the 33 stimuli (either from the web or repeating the pictures here) and again sum all your false alarms; that is the second data point of the first graph. Repeat for the third and final runs through for your third and fourth data point. For the second graph, repeat the process except adding up the ones you missed (Misses).

Looking at the graph of false alarms (see Figure 6-7), my subjects had an average of 4.28 of those errors on the first complete run/trial. That is, about 4 and 1/4th stimuli were mistakenly identified as being the target stimulus when it was not. If you inspect all the test stimuli throughout the pages, you can see that 12 of the pictures look nothing like the target stimulus. The Gibsons created 12 very dissimilar stimuli intentionally. They created the remaining 17 distracters to be easily confusable with one another. To do this, they generated a universe of spirals that varied along 3

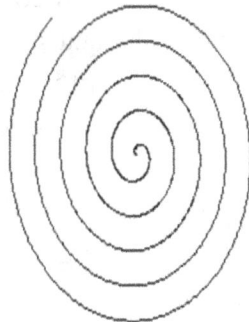

dimensions: number of loops, direction, and amount of compression. The number of loops could be either three, four, or five, the direction of the open part of the spiral could be either left or right, and width could be either skinny, medium or fat. Combining all the dimensions so that there is a spiral of every type produces the universe of 18 spirals (3 x 2 x 3). They chose the spiral that had four loops, faced left, and was of medium width as the target stimulus (as did we) while the remaining 17 spirals served as the confusable distracters. In the first run of my replication, subjects mistakenly identified approximately 4 and $1/4^{th}$ of these 17 distracters; they said "yes" when they should have said "no". Looking at the data as a function of trial number, you can see that the number of such false alarms decreased with practice. By fourth and last trial/run, fewer than 2 stimuli when averaged across all subjects were mistaken for the target. Not all subjects had fewer errors with practice and some subjects started with no errors at all. But on average, people got better with exposure to the pictures.

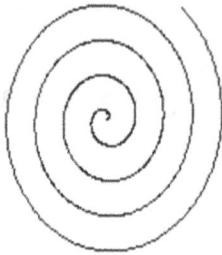

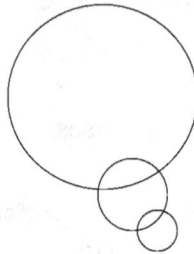

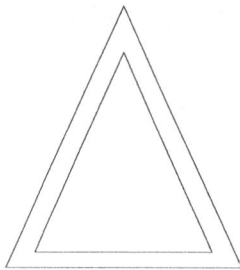

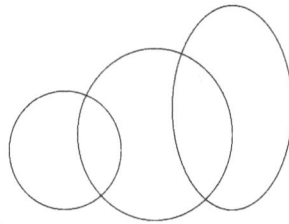

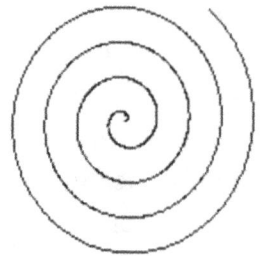

Before getting to the results on misses, consider other ways the data could be graphed. False alarms and misses are errors and one could graph successful performance instead: hits and correct rejections. If we graphed correct rejections instead of false alarms, the direction of the data line would go up rather than down. That is, people had more success with more practice. The two sets of graphs give equivalent information. Hits and misses are connected so picking either to graph reflects what happened in the experiment. Hits and misses add up to 4 on each run because there were 4 targets to be detected. Similarly, the false alarms and correct rejections are connected. They add up to 17 because there are 17 distracters that could be misidentified (not counting the 12 very dissimilar pictures that would not be confusable). Whether one chooses to pick the errors or successes to graph is largely a matter of personal preference. I prefer errors because perfect performance is always going to be 0 errors, which is easily spotted on any graph. False alarms and misses can also be placed on the same graph (or hits with correct rejections), although they could be of very different scale (that is, .2very different magnitude numbers).

Anyway, when we look at the false alarms, they decreased with practice as noted above. Thus, it looks like people were getting better at perceiving the stimuli with exposure to them. However, we need to look at misses, too, before drawing any conclusions. What if false alarms decrease, but misses increase? If false alarms decrease then subjects are saying "no" more often as a function of trial number when the stimulus is not there. But if misses increase, too, subjects are saying "no" more often when the stimulus is there as well. So combined, the person is just saying no

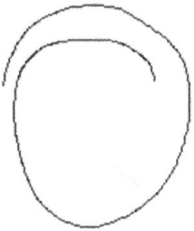

more often with practice, not getting better with practice. To assess improvement, the two types of errors need to be combined.

Here is a specific example that illustrates the point. Suppose there are just 4 pictures rather than 33 to make it simpler to analyze with the first 2 pictures being targets (correct answer is "yes") and the second 2 pictures serving as distracters (correct answer is "no"). On the first run through all the pictures, our imaginary subject always answers "yes" and on the second run through she always answers "no". (We will consider just 2 runs.)

		Run 1	Run 2
	Correct answer	Subject's answer	Subject's answer
Trial 1	Y	Y	N
Trial 2	Y	Y	N
Trial 3	N	Y	N
Trial 4	N	Y	N

When the subject always says "yes", she is clearly at chance—just guessing, you might say. With 2 possible answers, there is a 50-50 chance of being right with any answer. We would expect someone who is guessing will get it right about half the time. In run 1, our subject gets it right half the time. If we look at the false alarms, they decrease with practice. They went from 2 on run 1 to 0 on run 2. It is tempting to quickly conclude she has improved with practice. However, she is not getting any better at the task. Look at run 2, where she is answering "no" all the time. She is still at chance, again getting only half right. Note how the false alarm errors decreased from the first to the second run—but so did her hits. Graphing the errors would make it very clear that the misses increased from 0 to 2 at the same time as false alarms decreased from 2 to 0. She did not improve with practice. What *did* happen from one run to the next? There must have been some kind of change in strategy from always saying yes to always saying no, perhaps in an

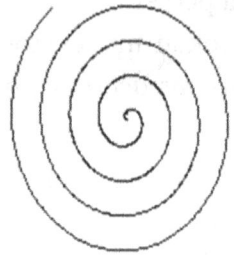

attempt to do better. By considering both false alarms and misses together, one can separate the effects of just strategy from real perceptual ability.

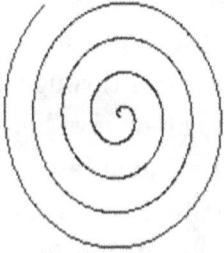

If you are having a feeling of déjà vu—a feeling incidentally that will be referred to in discussions of time perception, Chapter 10— it is because I discussed this earlier in the chapter. When O'Toole and colleagues had subjects determine if a picture was of a man or of a women, they converted subjects' performance into d'. Recall they did this using Signal Detection Theory (SDT) in order to separate out real differences in abilities from differences in strategies when judging different race pictures. As discussed, by neutralizing any effect of strategy, just the perceptual ability remains. The problem here has the identical logical format, with runs instead of race pictures, and yes and no responses instead of male and female responses. Yes/no responses are the general case and Signal Detection Theory was derived considering rates of false alarms and misses. In fact, O'Toole and colleagues, in order to use SDT, arbitrarily set the error of saying "female" when the picture is of a male to a false alarm and saying "male" when the picture is a female to a miss.

When I ask students to graph data, some of them have the intuition to combine both types of errors. This is especially true if I did not specifically direct them on which errors to graph. Usually, people who do this add the two errors together to get one measure of performance. You cannot just add false alarms and misses because the maximum possible number of correct "yes" responses is different than the maximum possible number of correct "no" responses. However, I love it when people have this intuition,

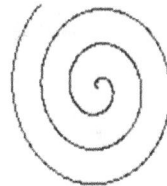

since the two types of errors do need to be considered together to figure out if there was improvement in the experiment.

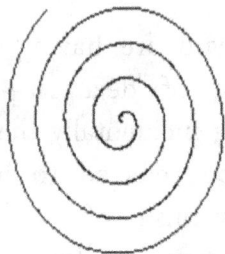

Going back to the data in Figure 6-7, it is apparent that misses also decreased with practice. Subjects fail to detect on average 1 of the 4 targets during the first run through all the stimuli. After 4 runs, they got better because they only miss on average less than half of what they did initially. Even without knowing the details of SDT, we can tell that our subjects did, in fact, increase their accuracy because both types of errors, false alarms and misses, decrease as a function of practice.

Out of curiosity, which type of error did subjects make more often? It looks at first as if that would be false alarms—more than 4 vs. only 1 miss in the first run through the pictures. Since there are only 4 possible misses though, missing 1 of them is 25%. That's pretty close to the false alarm rate: There are 17 possible realistic false alarms, not counting the dissimilar pictures that only the brain dead would mistake for the target. On the first run, subjects had 4.28 false alarms out of a possible 17, or 25.2 %. The two types of error rates then are pretty evenly matched. Anyone who thought to graph their data in percent has good instincts since false alarms and errors are easier to compare when using percents rather than absolute numbers.

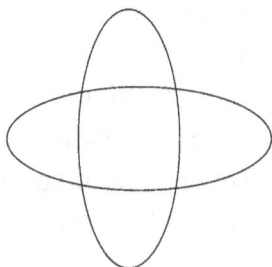

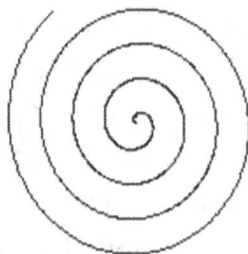

To digress just a tiny bit further, I also hope your instincts led you to use line graphs rather than bar graphs. Line graphs are a better choice here because there is a continuous dimension on the X-axis. Amount of practice is what is being considered and that varies continuously. Note how one can conceive of testing an intermediate amount of practice with a half a run through the pictures and the data might be expected to fall in between the data points that are there now. Line graphs make sense for continuous variables and bar graphs for discrete variables. Although likely no one will catch me on this, was it legitimate, then, to use a line graph for the Other Race Effect, such as in Figure 6-2? A bar graph is arguably a more appropriate choice for those data because "Asian pictures" and "Caucasian pictures" on the X-axis seem like separate discrete categories rather than a continuously varying dimension. But the visual crossover between the races when using a line graph is *so* compelling an indicator of the ORE, I found it hard to settle for a bar graph even if it was more appropriate. Perhaps Asian pictures and Caucasian pictures are really just two points on a continuum of faces with intermediate blended faces easily generated; if so, a line graph is a legitimate depiction of the data after all.

Ok, how do your data compare to the comparison data? How do the comparison data compare to the outcome of the Gibson and Gibson experiment from fifty years earlier? For the latter question, the Gibsons' subjects had a false alarm rate ("mean number of undifferentiated items on first trial", pp. 36) on the first run equal to 3.0. This is for adult subjects, as they also ran children. Instead of running everyone through 4 runs/trials as we did, they customized the length of the experiment to each individual such that reaching a false alarm rate of 0 in a run ended the experiment. It took an average of 3.1 runs to reach that point. To compare their data to ours, you can think of their false alarms as roughly equal to 0 on run/trial 3. If you are curious about how the kids did, older kids initially had 7.9 false alarms which took 4.7 runs to eliminate and younger kids had a

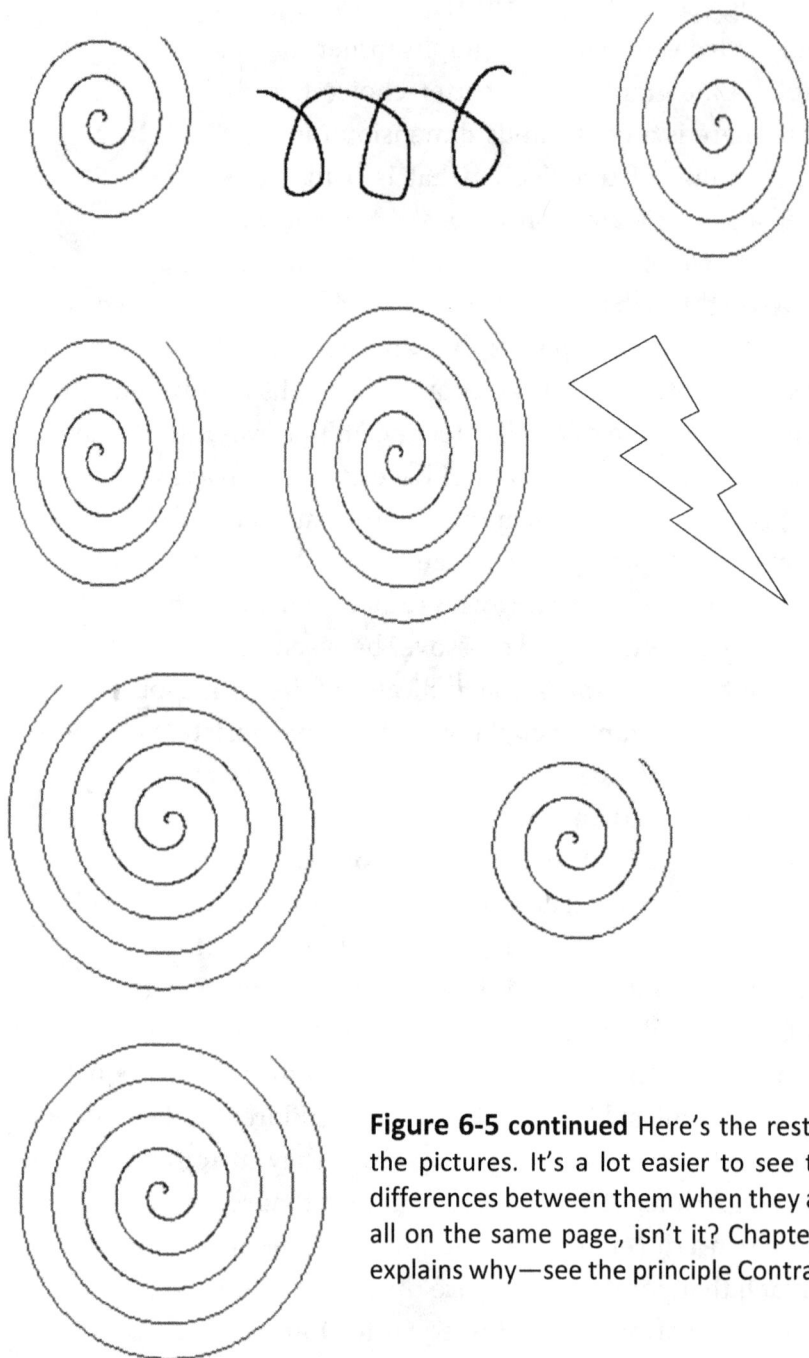

Figure 6-5 continued Here's the rest of the pictures. It's a lot easier to see the differences between them when they are all on the same page, isn't it? Chapter 7 explains why—see the principle Contrast.

whopping 13.4 false alarms at first. They ended up terminating the experiment early for many of these younger kids, since after 7 runs and lots of patience, they still had on average nearly 4 false alarms. In terms of the big picture, the replication produced generally comparable data: a few initial false alarm misidentifications in adults when shown a bunch of similar-looking spirals diminished with repeated exposure. We get better, too, like their subjects.

I will leave any further detailed comparison of results to you. If you find any differences between their data and ours (or yours), procedural differences could be to blame. Here, then, are details of their procedure. Their subjects were 12 adults—age and gender unspecified—the target item was studied by subjects initially for "about 5 seconds" (pp. 37) and the 33 pictures tested on each run was shown for 3 seconds each. In addition, each picture was an ink drawing printed on a card in their pre-personal computer era and the deck of cards physically shuffled between runs. The actual image they used for the training stimulus is a sketch of a spiral shape shown below (Figure 6-8). For the pictures that you judged in this chapter, I used an exact Archimedes spiral rather than a free-hand drawing and used computer-generated compression of 75%, 85%, and 95% for the three values along the fat/skinny dimension.

Figure 6-8 The spiral drawing used for training in the original Gibson and Gibson experiment. Gibson, J.J. & Gibson, E. (1955). Perceptual learning: differentiation or enrichment? *Psychological Review, 62*, 32-41. Published by the American Psychological Society; reprinted with permission.

Interestingly, the Gibsons do not report misses! It is unlikely that this is because not a single subject missed any of the targets in their study. Rather, they had an agenda to promote and it is only the false alarms that met their goal. They wanted to show that an object can easily be mistaken for a similar looking object until experience makes them look different. Yet we have seen that misses need to be considered together with the false alarms to ensure that an observer's perception really did improve. Otherwise, the decrease in false alarms could merely reflect a change in his or her strategy. Is this a flaw that reviewers should have detected and made them correct before going to print? Sometimes scientists who have an agenda that they are passionate about ignore anything that detracts from it. On the one hand, that is sloppy research, misleading even, and could lead to mistakes. On the other hand, it is those people that tend to change the accepted theories in a field and remain well-known for decades. It's an odd trade-off.

What Does It All Mean?

Improvement on perceiving spirals is another illustration of perceptual expertise. As a novice, all spirals look alike. You can easily distinguish a spiral from a triangle, but not from another spiral. Yet when you see them repeatedly, something changes in your perception. They now really look different. A true expert could never misidentify a three-loop spiral for a four-loop spiral because they no longer look anything alike. Don't care about nonsense spirals? No one does (though sometimes researchers forget). Had something you do care about been used for the experiment though, say taste of chocolate (we've done this!), some of the subjects might already be experts making it difficult to study how people become experts. By creating a brand new universe of spirals, everyone begins on a level playing field—only novices allowed. Moreover, that universe can be controlled allowing for careful manipulation of how many stimulus features differ and by how much. Fortunately, expertise on artificial spirals has a lot in common with expertise on natural faces, the focus of the last chapter and Part I of this chapter. Use one to learn about the other[6].

What were you aware of while you were a subject? Did you become aware at that the spirals were varying along three dimensions? Did they look more distinct as you kept looking at them? Introspection is always interesting and should be done, even though it is not the best window into what happens behind the scenes. Incidentally, I should mention that improvement at new tasks occurs for all sorts of reasons, not just perceptual expertise. The spiral discrimination experiment was not designed to rule out alternative reasons for improvement. We will have to live with that.

If you ever lose the intuition of what this change in perception is about, whether it be for spirals or faces or anything else, remind yourself of what the Gibsons conclude: "A stimulus starts out by being indistinguishable from a whole class of items...and ends up by being distinguishable from all of these."

Do all the collies that played Lassie on TV and in the movies look alike? Not if you own collies. Welcome to perceptual expertise.

Figure 7-1 Are you a perceptual sheep expert or a novice? Sheep just look like sheep to a novice but each individual looks different to an expert. Sheep races draw more than 100,000 visitors a year at The Big Sheep Entertainment Park in Devon, England. Photograph by owner, Rick Turner. Reprinted with permission.

Acquiring Perceptual Expertise in Seeing, Tasting, Smelling, Hearing, Feeling, and Immuning

Okay, so you've decided you want to become a perceptual expert. You want to dazzle your friends with your ability to identify beer while blindfolded. You seek to become a fancy radiologist with the ability to spot small cancers on medical scans. You decide you want to breed sheep for fun and profit and need to be able to distinguish Snowy from Dolly. Whatever your perceptual poison, help is on the way. But don't forget to marvel first—and often. Remember: We are talking here about actually making you taste things you could not before. Or see, smell, hear and feel things that you could not before. See Figures 7-1 (page opposite) and 7-2.

To turn yourself into a perceptual Superman or Superwoman, what should you do? You may already know. Think back to what you are already a perceptual expert in and try to recall what you did to get there. If you were too young at the time—like when you first learned the speech sounds of your language—then try to generate what you think training should be. See how many of the items that you come up with match those described below. While you are doing that,

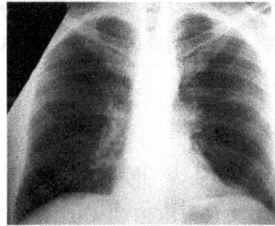

Figure 7-2 What would you say about this X-ray image? Novices have difficulty seeing much detail of interest.

let me apologize in advance. I am about to generate a list of seven items. Lists have a way of sapping the excitement out of even the most intrinsically interesting topics. Lists, though, can also be useful for organizing an area. Having a single set of explicit acquisition principles for this type of learning would be useful, I think. I gleaned these from assorted sources including the classic Gibson and Gibson article on improving the perception of spiral shapes (discussed in the last chapter), an old Eleanor Gibson book on perceptual discrimination learning[1] (which mercifully you do not have to read yourself) and some modern works, including a research article I will tell you about shortly. Incidentally, when referring to "acquisition principles of perceptual expertise", all I mean is how to get from being a naïve perceptual observer to becoming a perceptual expert.

Seven Rules for Becoming a Perceptual Superhero
Repetition

Did you come up repetition? In ordinary language, we are just talking about practice. Practice, practice, practice! To become a perceptual expert on an item, we may need to be exposed to that item over and over again. This seems like a lot of other learning processes. Practice makes perfect for any kind of expertise, perceptual or otherwise. Yet perhaps unlike other things you learn, acquisition of perceptual expertise sometimes can require unusually large amounts of repetition. Fabric merchants of old, who could identify cloth by touch, required years of apprenticeships for their well-respected ability. Training radiologists to read MRI images takes extensive specialty course work. Language comprehension, especially for a second language, can take years. Consider even tasks that seem simple like deciding whether or not two small line segments are pointing the same way or whether you felt just one or two stabs of a sharp object on your forearm. Can you tell those last two are laboratory tasks? They are called *orientation discrimination* and *two-point touch discrimination* respectively (see also Chapter 5). The names of the tasks are not important, but the tasks themselves are because they appear to take thousands of trials of repetitions to get really good at them.

We have a sense that practice is important but many unanswered questions remain. Is more repetition really needed to alter perception than to alter knowledge, like how to get to the mall or acquiring the meaning of the word "slacker"? How much repetition is needed and what is it exactly that has to be repeated? Do different types of perceptual expertise require radically different amounts of repetition? For instance, does it take more or less repetition to learn natural tasks like faces of another race than arbitrary stimuli like spiral shapes? How about becoming an expert at wines compared to fabrics or X-rays or hamsters or bird watching?

Contrast

Becoming an expert may well require gaining experience not just with one item, even if repeated over and over again, but with a whole array of different items within the area that you want to master. Let me try giving a couple of examples of what I mean. Everyone has heard of Pavlov and his salivating dogs. He repeatedly played a tone followed each time by placing meat power on a dog's tongue. Following such training, the tone by itself led to salivation. In addition, his dogs also salivated in anticipation of the food when tested with other similar tones they were not trained with—think tenor voices when trained on alto. Even after a thousand repetitions of the tone followed by the meat, the dogs still would not distinguish between different tones and would generalize their conditioned salivation to other similar tones. Interestingly, however, he found further that by adding to all this repetition just a handful of discrimination trials, and sometimes even just one discrimination trial, the results were very different. By adding discrimination trials, I mean that he added a second tone to training, similar to the one initially used, but followed by nothing rather than the meat. Now, the generalization previously seen was completely eliminated. The dogs only showed conditioned salivation to the specific tone that had been followed by food.

If you do not know much about Pavlovian conditioning, this example may be underwhelming. Also, we unfortunately do not know what the dog was perceiving when each of the tones was played. The experiments involved watching dogs' behavior rather than directly assessing their perception. Nonetheless the finding is still suggestive:

Massive repetition of just one auditory item did not reveal any ability to perceive it as different from another similar auditory item, whereas adding a second different tone—that is, *providing contrast*—made all the difference and led to clear discrimination between tones.

If that example does not do anything for you, you may be able to relate better to a real-world story that provides food for thought as well as tongues. Okay, here is the story. I used to eat Honey Maid graham crackers—a lot of Honey Maid graham crackers. I ate them practically every day for, I would say, about 4 and 1/2 years. I would have at least 2 crackers a day, sometimes 4, well alright, sometimes 8 crackers a day but in my defense, at the time they seemed a convenient snack that was low in sugar and trans fats. After all that time, one day the market was out of Honey Maid graham crackers so I just got a box of Nabisco graham crackers which they also stocked. They seemed fine, but when I was done with the box, I thought I had preferred the Honey Maid brand. Well, here is what happened. I went back to buying my original brand and while chewing the very first cracker of the very first box, I had this strong negative reaction and was pretty puzzled. I thought I could taste something that I never noticed before. I am not sure how to really describe the taste but it was not pleasant—I can still conjure it up vividly in my mind! It was a faint coconut-like taste but not in a good way and I could not ignore it. I could not bear it actually and every time I would eat the crackers I could taste something I did not like. I tried again the next day and the day after that. That was when my obsession with the perfect snack food came to an end. I stopped eating them.

I believe that what was going on was that even repetition of the Honey Maid graham crackers that I calculate to be in excess of 3285 crackers did not allow me to taste everything there was to be tasted in those crackers. Despite all that repetition, mere repetition was insufficient to turn me into a complete gustatory graham cracker expert. However, adding the contrast of the new brand of cracker proved to be very effective. For reasons I don't fully understand, when I tell this story to students, they find it quite engaging and even the usually non-analytical ones dissect it carefully and cleverly. One student asked how I know it wasn't just something wrong with that one new box of Honey Maid graham crackers. The point is well taken and the answer is that I did try again with another box and then about a month after that, I bought yet another box and I could still taste the aversive chemical. Whether it be tasting crackers, hearing tones, or seeing X-rays, perceptual expertise is aided by the contrast provided by experiencing similar, but different, instances in the class to be discriminated (Figure 7-3).

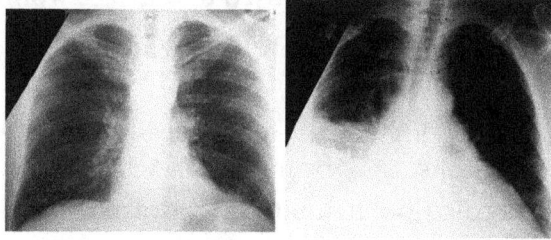

Figure 7-3 The X-ray from Figure 1 shown next to another X-ray. Does it help your perception to have a contrasting example?

Contrast is also a consideration when perceiving people of other races. If someone were adopted into a racial group different than his or her native race, we expect the person—let's call her Hope—to be better at recognizing people from the adopted race compared to people who do not have such close contact with the new race. This was discussed when analyzing perception of different-race faces and the Other Race Effect. But it raises the question of how Hope will do at her own race if she does not even see them at the supermarket, let alone the dinner table. There is something very disconcerting about thinking that an adopted person would be terrible at perceiving faces of her own native race. It was pointed out earlier, though, that anyone can just look in the mirror and the question was raised about whether getting to see just your own face would be sufficient to cause expertise at your own race. We may now be in a better position to address this issue, having considered both the principles of repetition and contrast.

Generalizing from the examples of contrast, seeing one's self in the mirror, even multiple times every single day, should *not* be sufficient to allow the perception of all the details needed to be a complete perceptual expert. This is because looking in the mirror provides just repetition of one item without any contrasting items. But the examples also suggest that the situation may not be as alarming as it first seems. Suppose after Hope sees her own face thousands of time, she glimpses just one more individual of her native race. That second contrasting item may now render all those trials of repetition meaningful, like it did with Pavlov's tones and Bedford's crackers. Hope may even be at an advantage over someone trying to learn new-race faces who was not exposed to one face from that race repeatedly.

The power of contrast was highlighted in this section by showing that the addition of one contrasting example following many repetitions of a single item can cause a sudden leap in expertise. We also expect contrast to be influential when perceivers are presented with evenly distributed multiple contrasting items, such as 10 different X-ray images, each repeated 10 times. Note I glossed over a bit the difference between expertise on a single item—the taste of one graham cracker—and expertise on a whole class of items from a related set or dimension—expertise on all brands of graham crackers. Consider also: Could I just study 2 different faces (or spirals or wines or crackers) and would that be enough to learn an entire race (or spirals or wines or crackers)? Two is certainly better than one, but still does not seem to be quite enough. Would getting a third example from the category allow even more perceptual details to be discerned? How about 6 different examples? When will the advantage of contrast reach maximum effectiveness? Is seeing 200 different faces just as effective as seeing 1000 different faces? Finally, how do contrast and repetition interact and trade off? For instance, would getting 50 trials of each of 2 faces be better or worse than 99 trials of 1 face and 1 trial of the other face? All of these questions are unanswered.

Attention
To continue building a list of acquisition principles, it turns out

that learning to perceive a new feature requires that you be intentionally attending to that particular feature, or at least to a task that use the feature. Rather than telling another anecdote, let me describe an elegant experiment to illustrate this principle. The experiment was conducted by psychologist Merav Ahissar at the Hebrew University in Jerusalem[2]. She uses the orientation discrimination task in which small line segments—in her version, white bars on a dark background—point in different directions. Detecting oriented lines is even duller than distinguishing among a universe of loopy spirals. But this does pare down perceptual expertise to its essence with just one tiny little perceptual difference along a single dimension—is it pointing this-a-way or that-a-way? This allows extraneous factors to be eliminated. So even if it's not a task you care about, you can extract the rules from such a pure case and then apply them to things you do care about, like other-race faces, medical scans, and wine.

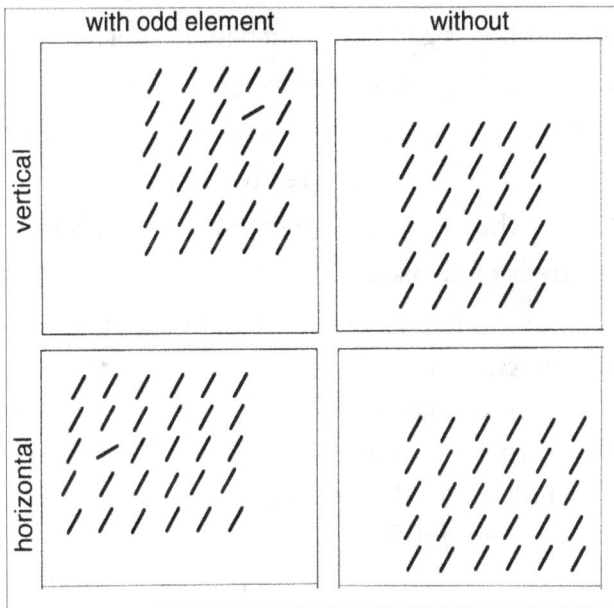

Figure 7-4 The 4 types of display in Ahissar's experiment. The experiment was an elegant demonstration of the importance of attention in reaching perceptual expertise. From Figure 2 in Ahissar, M. 1999. Perceptual learning. *Current Directions in Psychological Science, 8,* 124-128. Published by the American Psychological Association (APA); reprinted with permission.

In her experiment, participants could be shown any of 4 different displays, each consisting of many bar segments. There was a horizontal array that was 6 elements wide by 5 elements tall containing one bar oriented differently than the rest, the same horizontal array but with all the bars oriented the same way, a vertical 5-by-6 array with one bar oriented differently than the others, and finally, a vertical array with all bars oriented the same way (Figure 7-4). Every time a display was shown, half of the subjects had to indicate whether or not all the bars were oriented in the same direction. This is a difficult task because displays are only shown for $1/20^{th}$ of a second (!) and if a misoriented bar does appear, it can be in any position in the array. The other half of the subjects instead had to indicate whether the display was in the horizontal 6-by-5 arrangement or the vertical 5-by-6 arrangement, also a difficult task. Critically, as hundreds of trials progressed, all subjects got better and better at correctly making their assigned judgment, as we would expect given we know that repetition leads to perceptual expertise. However, when they were tested on the other task that they did not have to pay attention to, they were terrible! That is, they were no better than naïve subjects who never saw the hundreds of displays.

If you are not impressed yet, contemplate that all of the subjects *saw exactly the same displays.* The first half of the subjects could ignore whether the displays were horizontal or vertical because it was not relevant for their orientation judgment whereas the second half of subjects could ignore the orientation of the individual lines because it was not relevant for their horizontal-vertical judgment. In other words, if you are not paying attention to a feature then you do not improve at detecting that feature even if it is presented to you over and over again. I like thinking about this experiment because of the clever way that Ahissar holds everything constant between the groups —what they are shown, how long they get to see each display, how many trials they get—except for what the subjects themselves attend to. She manipulates attention and in doing so shows that attention is required to obtain perceptual expertise.

Here too, we have unanswered questions. Attention is known to be a fuzzy concept. What does it mean to "pay attention"? Is attention just another way to say we need to single out the relevant feature? (See next item.) Is the need for attention to a feature to acquire superior detection ability the same as saying we need to be consciously aware of a feature? There is something circular about the relation between perceptual expertise and attention, but I will omit it because it is too abstract for the present "how to" section.

Single out a relevant feature or dimension

Acquisition also seems to be facilitated by any procedure which helps the observer know what to look for. That is, you learn faster with a procedure that isolates in some way the property that distinguishes one item from another. Example? If you want people to be able to see the difference between airplanes, you might say "look at their noses", a procedure that was actually followed during wartime and funded by the military. Suppose for the spiral discrimination experiment you tried in the last chapter, I had said: "Try counting the number of loops and while you're at it, look at the width and where the spiral's opening is". You would now be able to see the differences between the similar-looking spirals much faster than had you not been given the hints about the 3 dimensions along which the spirals vary. Just talking to you can improve your actual perceptual ability!

This fact may have been exploited to the max in an experiment that was conducted on *chicken sexing* of all things. Hatcheries need to determine whether very young chicks are male or female to not waste resources on non-egg laying males. If you were to try to sex chickens—see Figure 7-5—you would be only a little better than chance, but experts can do so very accurately. Chicken-sexers are in very high demand and the expertise often takes years of training and thousands of chicks of practice. Perception psychologists Irvin Biederman and Margaret Shiffrar turned college students and faculty into experts immediately[3]. They essentially instructed them on where to look (about 2/3rds to 3/4ths of the way down in between or slightly above a gap formed by 2 horizontal cylinder shapes at the edges of the photo) and what to look for (a round "bead" is male and either flat, or pointy like a wide V, is female). By singling out the relevant items, naïve observers now performed at a level comparable to seasoned experts. Otherwise, little chicks, like all natural objects, are so complex that no one knew where to look. Try again with Figure 7-5 and see if the additional information makes the task easier.

Figure 7-5 Take your best guess about the sex of each of these one-day-old baby chicks (top row) from photos of their undersides. Chicken sexers with years of practice are very important perceptual experts at hatcheries. Top: Adapted from Biederman, I. & Shiffrar, M.M. (1987) Sexing day-old chicks: A case study and expert systems analysis of a difficult perceptual-learning task. *Journal of Experimental Psychology: Learning, Memory, and Cognition, Vol 13(4)* , 640-645. Published by American Psychological Association (APA); reprinted with permission Bottom: Chick with eggs, http://hubil.free.fr/Wallpapers/1600x1200/Animals/Animals%20-%20Babies/

Verbal descriptions are not the only way to single out a relevant property from all the background noise. Arrows or boxes or darker ink can point to an informative region (Figure 7-6). Ever have a caricature drawn of yourself at a party? These emphasize some features artistically by exaggerating them, making them very hard to ignore (Figure 7-7). You can also accomplish the same goal by de-

Figure 7-6 See if it is easier now to tell apart two of the sheep's faces from Figure 7-2. Look between the arrows for one region to inspect closely.

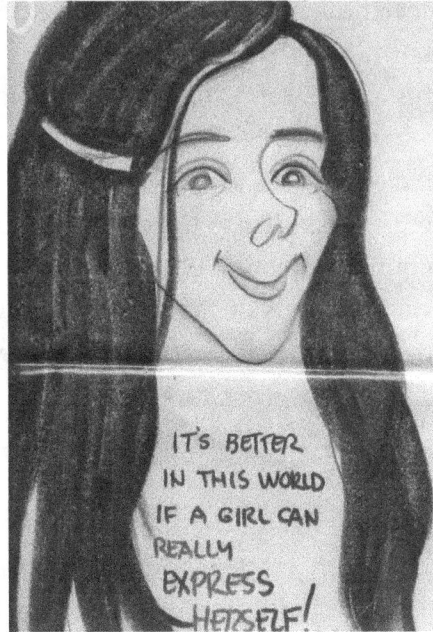

Figure 7-7 Artistic caricatures exaggerate features which draws attention to them. Pop singer Justen Bieber (big hair/head…), country singer Carrie Underwood (radiant smile), & your author Felice Bedford, at age 14 (pointy chin). Bieber by Dick Kulpa starletteuniverse.com Underwood by Cathy S.R. Reed cathyreadart.com Reprinted with permission. Bedford drawn at Brian Dresner's Bar Mitzvah, artist unknown.

emphasizing the remainder of the display. For example, if you want to determine which of two medium size eggplants is lighter so as to choose the one without seeds, what do you do? Instinctively, you shut your eyes as you consider their heft. This blocks out irrelevant information allowing the perceptual system to process only the relevant differences for the task at hand.

Knowing which features are relevant to perform a task is likely responsible for a finding Ahissar labeled the Eureka effect. Those orientation discrimination tasks she used can take thousands of trials before subjects can accurately see whether or not a display contains a misoriented bar. But she also found that if subjects first had a few trials where they could look at the display with a misoriented bar for a long time, rather than the usual fraction of a second, then they very quickly became experts. Extra time permits the exploration needed to see and inspect what a misoriented bar looks like. You then at least know what you are looking for and now have a fighting chance of finding it even when the task is made faster. Before, you didn't know what "it" looked like. Finding a needle in a haystack is easier if you have at least seen what a needle looks like. Note how the Eureka effect is similar to the chicken sexing ability. When confronted with complex stimuli consisting of multiple irrelevant dimensions of variation, advanced knowledge saves thousands of trials of having to figure it out the slow way, one tiny detail at a time.

Reward and feedback

Feedback is a very intuitive concept. If we find out whether we are right or wrong then we know what to do for next time. We learn. However, in perceptual expertise, feedback is unnecessary. Even without finding out if a perceptual judgment is correct, improvement nonetheless occurs from repetition, contrasting examples, attention and singling out informative features—the other principles of acquisition we have been generating. Moreover, when feedback does appear to be influential, its influence is unusual. Cognition researchers Ling-Po Shiu and Hal—used to go by Harold before moving to

California—Paschler found that feedback improved performance in an orientation discrimination task even though it was given after a whole block of trials rather than after each trial[4]. That's a very peculiar kind of feedback. Suppose you were doing math problems and after doing 40 problems, you were informed that you got 70%, or 28 answers right. Wouldn't you need to know *which* ones you had gotten right to learn from the feedback? In perceptual expertise, the role of feedback may be to let the perceiver know if she is attending to the right feature. If the feedback is positive, she is on the right track and if it is negative, she may want try attending to a different feature in the future. Feedback, then, may just be another method for implementing the last item discussed, attention to the relevant dimension, in this case by playing a game of "hot" and "cold" rather than directly telling the person what to attend to.

Many of the unanswered questions here are more theoretical rather than relevant to a practical how-to training course and some will be considered in a later theoretical chapter (Chapter 12). These include: Can we relate this back to the feedback we considered in Chapter 3 on variable and constant error and estimating the sizes of objects through touch? Is feedback also dispensable in other kinds of perceptual learning? How does feedback relate to error detection? Can feedback be entirely internal rather than provided by external sources?

Easy to hard

Let's say someone wants to be an expert in achromatic colors; those are the boring neutral ones ranging from white to black that can be worn with any outfit. To make someone an expert in perceiving subtle shades of grey (no, I am not referring to the bestseller *Fifty Shades of Grey*!), start by making the difference between 2 grays very large and then, as expertise is acquired, make the difference smaller and smaller. Pavlov discovered this principle early in his conditioning research and it has been rediscovered by perceptual expertise experts. Whether it is for a distinguishing within a single dimension, like amount of grey color or amount of line orientation, or for more complex multi-dimensional items like faces, wines, or baby chickens, perceptual expertise is created by starting with (contrasting) examples that differ greatly from one another and then increasing the difficulty of the discrimination only after the easy ones are mastered. This, too, is very intuitive. No one teaches calculus before addition. Whether there is something different about progressing from easy to hard examples in perceptual expertise than in other learning processes is not known, but for a training program it does not matter. Throw everything we have at the learner in an attempt to improve perceptual detection as much and as quickly as is possible.

Labeling

Normal Lung   Pneumonia

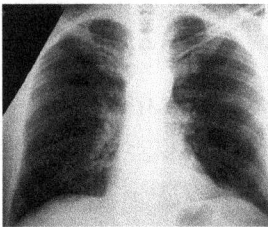 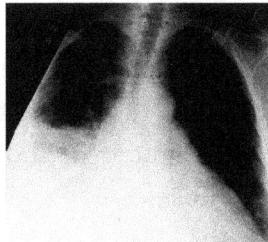

Figure 7-8 Labeling items to be distinguished can facilitate acquisition.

Give it a name! Would it help to read the X-rays from Figures 7-2 and 7-3 if they were labeled "normal" and "pneumonia" (see Figure 7-8)? Research on memorizing faces suggests that you are more likely to recognize pictures of faces if you studied them with labels—Bob, Tom—than without. Seems like some kind of magical placebo effect, but we appear to be hardwired to use names with perceptual distinctions. Psycholinguist Ellen Markman at Stanford University showed that preschool children assume that each new word refers to a unique object[5]. If you point and say "that's a blicket" and point to the same place again and say "that's a dax", the children infer dax must refer to a different object or property rather than concluding blicket and dax are alternative names for the same thing. I think that this assumption is deeply related to error detection in perceptual learning but I keep reintroducing error detection—the key concept I argue in all of perceptual learning—only to snatch it back. It belongs in the later theory chapter. Practically speaking, to facilitate becoming a perceptual expert, give each example you study a unique name.

Otherwise, We Do Not Perceive Very Much

And there you have it. Seven ways to achieve super-human perceptual abilities. You are well on your way to becoming a perceptual expert in any area you so choose. To step back and regain perspective on the big picture, we are analyzing a learning process whereby experience causes you to see more than you did before or hear more than you did before, and so on through all the senses (for a new sense modality—immuning—see Chapter 11).

Without such perceptual learning, you do not take in everything there is to be perceived. Remember when you recalled the features of a penny from memory or more accurately, when you were not able to do so? If that did not impress you, consider additional examples that illustrate that you just do not perceive everything there is to be perceived without special training. Then we will consider its significance.

Look at the two similar pictures of my class in Figure 7-9 (the following 2-page spread) and try to find the differences between them. Finding differences is similar to search puzzles like finding Waldo. When I present this task to students and subjects and then time their searches, quite a few people comment that they thought it was going to be easy but then find otherwise. Search puzzles are fun because we seem to be absolutely mesmerized when led to believe there is something in front of us to be seen but we can't see it. How come I don't see it? We get drawn into searching and searching so that we can experience it for ourselves and then once Waldo is spotted, or the differences in the photos detected, then it looks so completely obvious. I argue that this belief that our perception could not possibly fail us—no matter how many times we experience that it does—is really very strong and is, in fact, so strong that people invent spirits.

Figure 7-9 This is my Happy Class (I am in the front row, second from the left). Can you find five differences between this photo and the one on the next page (plus a bonus very hard to find sixth item)?

Figure 7-9 continued We time people and record their thoughts and percepts while searching. Answers at the end in Notes[*]. Happy Class photo by Whitney McNeil, alteration by Laura P. Garcia and Felice Bedford

To explain: Do you ever have the experience where you look for something, can't find it, look everywhere, still can't find it and then later discover it was right in front of your face the whole time? This happens to a lot of people, most of us I would maintain, but it is not always attributed to the way perception works. "Well, grandma must have taken it and put it back; that was her favorite pair of earrings". The idea that long-departed spirits take things and put them back is tempting when there does not seem to be any other explanation. What is really happening, though, is that your perception is capable of far greater tricks. Well, it would be a great trick, too, if grandma could return as a thief, but it is your perceptual system to credit or blame.

This is actually the second example mentioned where other-worldly causes are attributed to perceptual events because we do not really understand how perception works. When I say "we", I mean here people in everyday life who do not give a moment's thought to perception, instead assuming that it is stable, flawless and permits all worldly occurrences to be detected. Therefore, the explanation for an odd perceptual experience is assumed to lie elsewhere. The other example was astral projection, in which the soul leaves body and goes traveling. The overlooked perceptual principle was body schema, including an inference on the location from where we are doing the perceiving. For both examples, it is perception at work, not spirits, although I do think spirits break up pairs of socks.

Perhaps the most compelling example of failing to perceive all the objects that are out there is known as change blindness or inattention blindness[6]. You are giving directions to a stranger when he bends down under a bench to tie his shoe. When he stands up, he has been replaced by a different guy—*but you don't notice.* A woman with an umbrella walks through a basketball game while you are trying to count the ball-passes for one team and you don't see her. Surely if she were a cheerleader you would, except, that you don't. Not even a gorilla gets your attention. You fail to notice something big has changed in your environment and then when you are shown it later, it is so major that you just cannot believe that you missed that! Change blindness tends to occur when engrossed in a task like monitoring the performance of a sports team. It also occurs if vision is briefly interrupted, like when someone disappears behind a bench. Experiments on change blindness were first published by Dick Neisser and Robert Becklen decades ago. In recent years, the phenomenon has been popularized by others with each example becoming more spectacular and incredulous than the last. Life's scenes are complex and change blindness further shows that at any one time, we cannot perceive everything that is out there.

You do not see everything. The learning process of perceptual expertise enables you to perceive things that you did not before training if it is important to your life to do so. If there are things out there to be seen and you are capable of learning to see them, why wouldn't you just see them to start with? Isn't that simpler? The issue of why on earth something seemingly less straightforward would occur instead was addressed during discussion of the Other Race Effect. The answer applies to all examples of perceptual expertise and bears repeating. There is too much information to possibly perceive it all. Moreover, human machinery is not the real limiting factor. Rather, it is that you really would not want to perceive it all, even if you could. Change blindness suggests that successfully concentrating on one thing means needing to ignore everything else. But it is a lot more than that. It would be very hard to function if you saw every detail. Sleeping Tony. Standing-up Tony. Tony from the left. Tony from above. Tilted-head Tony. If you let through all Tonys that differ each time we glimpse him, you would introduce yourself to the same person over and over again. It is not very adaptive to act like there are 6 different people when there is only 1person. Unless you categorize all those different views together and fail to take note of the differences, you are going to behave inappropriately.

You learn as a toddler what a chair is for, a good thing even for an energetic youngster. Suppose you now notice the differences between all chairs rather than their similarities. What if every single chair looks different to you, so different that you did not categorize them as chairs and conclude that each one is just a completely different object? Then you would have to learn for each and every chair what it is that you can do with it. To say that doing so would be extremely inefficient would be an understatement. Survival itself is questionable if you could not generalize learned behavior to similar objects. If everything really looked different to you, you would have a separate category for each and every percept and that would require you to learn anew each behavior that was appropriate for that particular item. We cannot function like that. We have to be able to generalize. We have to be able to learn what behavior is appropriate and sensible for one thing and then generalize to the things which appear similar. Seeing the differences between similar items is just going to get in the way of being able to categorize them and generalize.

To elaborate further, when you learn to discriminate between similar looking things, it is the opposite of being able to generalize. The default cognitive state without training is typically to generalize between similar items and not notice the fact that they are different. This serves you well because you do not have to learn the same thing repeatedly. However, something may happen in your environment that causes you to need to be able to detect the difference between two similar items—one berry is poisonous, the other is safe, one sheep is yours, the other belongs to your neighbor. Acquiring perceptual expertise can then enable to you to discriminate; you no longer generalize because now two things actually look different and you can behave differently towards them—eat only the safe berry and collect only your sheep. Typically, you do not have both. You either have big categories and generalize within the category, or you become a perceptual expert with smaller categories and can behave differently towards each item.

Here is where it starts to get important. Note how sometimes you need to see differences, such as when you try to distinguish among different people that you encounter in your environment. It would be catastrophic not to. But sometimes you also need to *not* see differences, such between Tony when you view him from the left and Tony when you view him from the right. He is still Tony and to see the different images otherwise would also be catastrophic. *You need to both see difference and not see differences.* What you need depends upon the situation and what you are doing in your life at that moment. A sheep herder needs to distinguish one sheep from another. The rest of us are better served by having one big sheep category and generalizing what we learn about sheep to the whole class of sheep as distinguished from what we know about, say, wolves. It's a sheep, not a wolf. We need the perceptual expertise mechanism to allow us to see differences when the need arises and to (hopefully) lose the ability to see differences when it does not serves us well. There is no right answer to the question of whether two things are different or not. The two things are both different and not different depending on the exact use. It is also more perceptually efficient to detect only what is needed, although that may be beside the point. It is this balance, this titration back and forth between seeing and not seeing, which is critical in the process of perception—and what has been missing in the study of perception.

To digress briefly, it is certainly not new to point out that generalization and discrimination are opposites or that generalization is often the default state before special training. No one dissected this better than researchers of Pavlovian conditioning and it has since been rediscovered countless times in the fields of learning. But in the field of perception, there has been a failure to recognize the importance of the continual see-saw between generalization and discrimination. *Categorization* (generalization) is discussed by a few researchers, but it is considered a topic of only marginal relevance. Instead, a century of my perception colleagues at large have missed the boat. The big focus instead has been on *object perception*. What this does is to assume that what we perceive is static. We assume that there is a thing out there, an "object", and ask how we perceive it. This approach fails to appreciate that what there is to be perceived depends on the situation and on us; it has to change all of the time. Does this sound familiar? It is quite similar to the mistake the "uneducated" perceiver makes by assuming his/her own perception is stable. But it is less forgivable to those actually studying perception rather than just using it. No wonder in one hundred years there has been little satisfying progress or solutions to the sub-discipline within perception of object perception. What about modern trends and discoveries? Yes, there is a modern trend. Fueled by the technological ability to scan our brains while we perceive, rather than ask how we perceive an object, now it is fashionable to ask: How does the *brain* perceive an object? New technology, same mistake. (Tip number something— I have lost count—Don't assume that study of the brain will provide all the answers.)

An Exception: Autism

Autism is diagnosed in 1 out of every 88 births. The severe form brings with it profound deficits in language, social interaction, and cognition. But there are also intriguing perceptual characteristics. The autistic individual has been reported to be unable to filter out irrelevant detail[7]. Might he or she be experiencing a world very much like the one described earlier, in which every stimulus is interpreted as something new? If no detail is ever missed then far from being a gift, this would be an affliction at least as bad as an inability to speak. Seeing everything there is to be seen would be catastrophic, as discussed earlier. In ordinary perceivers, the default perceptual mode before special training allows for generalization to similar items. In autism, this mode may be replaced by an ability to discriminate between everything that is different—which is everything—without needing any training. Everything! It would be easy to see then why moving an item of furniture to a new location would be so upsetting to an autistic child. If all details are noticed then one item of furniture in a new location is not one item of furniture in a new location— that's what it is to you who has categories. To someone with autism, it is an entirely new visual array, unrelated to the first and therefore probably intensely scary,. What would it feel like if you returned home to find every item of furniture dissembled and scattered in different places. One might imagine an autistic sufferer's painstaking attempt to achieve and hang on to stability in the face of all those ever-changing perceptual details that occur even with the simplest head movement. And then having it thwarted by a physical change to the environment that you would not even notice—a toy soldier moved one inch to the left. The environment had to be memorized as looking exactly a certain way under each of a thousand different views and now requires a whole new sensory array be deciphered. What little stability an autistic has been able to extract would be easy to ruin.

If it is the case that autistic observers perceive every detail then they should do better than ordinary observers at search puzzles, like finding Waldo, and in change blindness situations. Indeed, it has been found that autistic subjects are able to detect more of the unexpected changing events than ordinary observers in change blindness paradigms. If autistics are essentially natural born perceptual experts before any special training, does this mean they can more easily achieve perceptual expertise in some of the common areas in which ordinary observers seek training, like reading X-rays? I do not know of any work that explicitly addresses this question but unusual, seemingly perceptual, talents have certainly been reported in individuals with autism. One example is British musician and autistic Derek Paravicini who has a remarkable ability to reproduce any music on a piano he hears but once. Whether all sense modalities are equally affected in autism is also an issue that warrants scrutiny. If even one perceptual modality is spared the onslaught of processing every detail, olfaction say, then it may well be an autistic's pathway into the normal world. Finally, it is also worth investigating whether the core of autism, that is, the primary problem from which all the others derive, is social, cognitive, linguistic—or perceptual. Autism may belong in our Chapter 1, perceptual deficits.

Abstract Discussion

Is perceptual expertise a type of perceptual learning? Recall that while the general term "perceptual learning" has historically been co-opted to mean perceptual expertise, I have offered a different definition of perceptual learning while using the term "perceptual expertise" to refer to this Gibsonian perceptual differentiation process. We can consider both the general definition of perceptual learning and the t1-t2 framework to assess how perceptual expertise measures up. According to the definition, any change in actual perception that is brought about by experience and removes an error in perception, or otherwise improves perception, is a candidate for perceptual learning. For the framework, a stimulus at time 1 produces a certain percept and following experience, the exact same stimulus at time 2 produces a different percept. Experience has produced a change in actual perception.

Starting with the framework, consider its application to stimuli from the Gibsons' universe of similar-looking spirals. One spiral is fat with three loops that face left while another also is three-looped and left-facing but is only of medium width. Perception at t1 as a novice is that the two stimuli just look the same. They are perceived as spirals, distinguishable from triangles or squares, but they both just look like spirals. Some experience is provided—I will come back to that—and at t2, the same two spirals are shown. Now, no longer a novice, one spiral is so clearly seen as fatter than the other. The expert can actually see that there is a difference. They no longer look the same. Thus, we have exactly the same stimuli, the pair of spirals, that produce different perception before and after experience. Initially the spirals produce one percept—look identical—but later, as a result of experience, produce a different percept—look different. What is the experience? The experience is comprised of all the acquisition principles that we have been discussing. In a nutshell, the experience consists of paying close attention to the relevant features of contrasting, preferably named, examples repeated frequently. Perceptual expertise does meet the perceptual learning framework.

Now for the definition. I think we have already established that perceptual expertise reflects a change in perception brought about by experience. The second half of the definition is the tricky half: "...removes an error in perception or otherwise improves perception". I argue that there is a hidden error in the perceptual expertise process. I say hidden because perceptual expertise has never been discussed as involving errors. This is unlike the other major perceptual learning process, perceptual adaptation, which, as will be seen in later chapters, has always involved detection of errors or discrepancies. *Error detection is key to understanding all of perceptual learning, including the hidden error in perceptual expertise.* Why is the detection of an error so important and what is it here for this process? I will postpone discussion until we have had a chance to consider errors in the other learning process where it has been more transparent.

We can, however, consider now errors that we have already discussed and connect them to perceptual expertise. This line of discussion is quite abstract, so I will try to keep it brief. Earlier, two types of errors were introduced along with the task of estimating the height and width of objects thorough touch. They were variable and constant error, and mathematical equations were provided to describe each error. I also mentioned that, as we would see later, these two errors correspond to the two major types of perceptual learning. It is now later. You are in a position to figure out whether it is variable error or constant error that captures the perceptual expertise kind of perceptual learning, rather than taking my word for it. The answer is variable error. A decrease in variable error corresponds to the sharpening of perceptual ability that occurs with expertise. Recall that variable error is a measure of how precise perception is. Bob, who starts out with a high variable error, judges the same item differently each time—a response of "3" one time, maybe "12"another time. People with a high variable error are not being very precise. They just perceive one giant category and something which is a "3" does not look any different than something which is a "12". Instead, they are all grouped together. As variable error decreases, the categories get smaller and smaller and the percepts become more and more precise. If terms from statistics are familiar, you will recognize the "variance". If not, you will still recognize that this also describes perceptual expertise. In perceptual expertise, too, the categories subdivide and become smaller as the ability to discriminate among members of a big category gets better. One big category of white wine is recognized by novices but reisling, gewurtztraminer, and pinot blanc by experts. The same is true for spiral connoisseurs with fat loops, opening position, and loop repetition to a novice's "spiral". Regardless of the content area, the increased perceptual refinement that occurs with expertise can be captured mathematically by the decrease in calculated variable error.

Practically Speaking

A class in middle school to eliminate the Other Race Effect? Less time to become a radiologist? Training to sniff out cancer cells? These now all seem possible.

I would like to see a perception treatment developed for neuropathy of the extremities, often found in diabetics. I would recommend a training program that applies the two-point touch discrimination procedure to the bottom of the feet. Earlier, it was noted that improvement in two-point discrimination takes thousands of trials. But we now know that our training principles can speed up considerably the detection of anything. Improving touch perception in the feet for those who are losing sensation would improve walking and quality of life, and decrease the risk of falls and injury.

Dogs are great sniffers with one hundred times the number of olfactory receptors. They excitedly sniff out hidden wonders with every step. Are you really content to let them have all the fun? In the dog world, we would be considered handicapped, downright pitied, I am sure. A genetic disadvantage does not mean there cannot be substantial improvement. (A great example of this is improvement found in bitter taste perception following training among people with a common genetic bitter taste deficiency. The perception world seemed surprised by the improvement[8], but you and I know better.) Despite our clear need for practice in olfaction, we seem quite intent on doing the opposite with daily showering and heavy chemical scented products. How sad that we have given up the opportunity to become experts within olfaction. Some dogs can be trained to sniff out cancers. Can people? How would we train a person to be so good at olfaction as to detect whether or a skin cell is cancerous or not? By using the acquisition principles of perceptual expertise.

Software like *Rosetta Stone* that promises to speed up language learning is appealing because so many adults struggle for years with the acquisition of a second language. As far as I know, no one has tried using these seven principles to develop an efficient training program for second language learning.

We have, however, begun a program in my lab to apply these perceptual expertise acquisition principles to perceive faces of another race better than we do now. This brings us back full circle to the problem discussed at the beginning of this long three-chapter discussion of perceptual expertise, namely, the Other Race Effect (ORE). If it is the case that the Other Race Effect is an example of perceptual expertise then we ought to be able to use the seven principles gleaned from other areas of expertise and find them to be effective in the ORE as well. When I say "if it is the case", I am trying to re-emphasize that only recently has the ORE been described in the context of perceptual expertise. Of course, it has been recognized for a very long time that experience has something to do with recognition. Consider this century-old quote[9]:

"....other things being equal, individuals of a given race are distinguishable from each other in proportion to our familiarity, to our contact with the race as a whole. Thus, to the uninitiated American all Asiatics look alike, while to the Asiatic, all White men look alike."

It took nearly one hundred years before explicit connection was made to the general expertise learning process that occurs in all areas of perception, whether it be perceiving wines, chickens, or faces. Even now, the Gibsonian perceptual expertise process may get mentioned in the introduction of a research paper, but then dropped and its impact unanalyzed. We have begun to explicitly apply the perceptual expertise acquisition principles to the challenge of recognizing faces of strangers from other races.

Thus far, the approach has been successful. In one study, we used the principles of attention and singling out the relevant feature[10]. One group of subjects were instructed to attend to the eyes and told, in

A subject studied this side-view.

Then was asked if it matched any of the 6 front views above.

Figure 7-10 This is one of the faces that subjects had to look at and then try to pick out of a line-up of 6 faces shown one at a time. All photos are mug shots obtained from Maricopa County in Arizona. Subjects were also given the analogous task with spiral shapes adapted from the Gibsons' universe of spirals.

order to motivate them to do so, "studies have shown that looking at the eyes can help you recognize new faces". Another group of subjects were instead instructed to look for how fat spirals are and that doing so helps people distinguish between spirals. That is, everybody in the experiment gets to study both faces and spirals, but half the subjects get only the instructions on faces (inspect the eyes) while the other half get only instructions on spirals (inspect the fatness). We then looked to see how people did at recognizing the faces and spirals. One of the face stimuli used in the study is shown in Figure 7-10.

We found that, sure enough, subjects given instructions on the spirals learn spirals faster than those who do not get the spiral instructions but are no faster at learning faces. And critically: Subjects who got the instructions for studying eyes did better on recognizing faces that a previous experiment had shown were difficult to recognize. Overall, people with spiral instructions got better on spirals but not faces and people with face instructions got better on faces but not spirals. We think that this is a very nice effect because it shows that perception of faces is behaving just like perception of spirals. The experiment confirms that the same acquisition principles, such as attention and singling out relevant feature, do indeed characterize perceptual expertise in general, regardless of the subject area to which they are applied.

Seeing, Hearing, Feeling, Tasting and Smelling

I want to make sure there is a clear example of perceptual expertise for each perceptual modality. I know of no better way than to take you the experts themselves. Following are examples from 22 people in their own words that we collected in the year 2012. There are perceptual superheroes among us. They have developed detection abilities as good as any Spidey Sense.

Vision

Ever since I could remember, I've gone hunting with my father. Due to my extensive search for animals, my vision for spotting hawks, deer, javelina, etc. is better than most people. I can detect the smaller movements of animals far away.

I am a perceptual expert in figure skating. I grew up ice skating so I can instantly tell the difference between one movement from another, specifically jumps. To someone who doesn't know the sport, a jump is a jump. They all look the same. People who don't know the sport just see the skater jump into the air and land, where as I can tell the difference in technique with each jump. For example, I can see whether the skater was using their edge of the blade to take off or their toe pic. I can tell which direction they were taking off for the jump in and which leg they used to jump off of. Unless you know the sport very well, it is very difficult to see and be able to tell the differences between each jump.

I am a perceptual expert in distinguishing a weed (plant) called Leafy Spurge from all other plants. The weed itself can be confused from afar with plants like mustard and other yellow flowering plants. As far as I know, out of many family and friends I am the only one who can see spurge (or not spurge) and make the correct identification immediately.

I've watched heartbeats for the last eight years so i can identify several different defects with a person's heart rhythm rather easily at this point. I can't necessarily tell you why they're messed up but i can identify defects on an expert level.

I have worked at *Old Navy* (clothing store) for five years and I can spot their clothes on people I see while on campus from other clothing brands people wear. I do this just by how the shirt looks (where the shirt is cut) and the color since Old Navy has a specific range of colors for its clothes.

Hearing
I played a lot of basketball in the gym growing up. I feel that I can sense about how many people are in a gym based on the squeaking of shoes on the floor. Even though feet may hit the ground at very similar times I can get an idea of how many people are actually playing on the court by hearing the squeaking.

I can pick out different musical instruments in a band whenever I hear one. This is probably due to the fact that I've been in band for eleven years.

I have worked with children for eight years since I helped take care of my niece and nephew and proceeded with a job in daycare with toddlers. I can listen to children speak words and figure out what they're trying to communicate with ease. Not many people can understand toddler or baby language. I, however, with experience can fine tune words and gestures.

I would say I am an expert in distinguishing different dialects of Arabic when spoken because I grew up learning the language.

I have a one-year-old daughter, and I feel that I developed the ability to distinguish between my daughters cries: if she is tired, if she is hurt, if she is bored, if she is thirsty, if she is just being cranky.

Touch
I work in a sporting good store that prides itself in quality of fabrics for performance like breathability and warmth. I can feel a fabric and by touch alone determine if it is polyester, cotton, organic cotton, spandex, elastene, nylon, ripstop nylon etc.

I can recognize how many ounces a boxing glove is by putting it on. Sizes vary depending on fighting gloves (usually ten ounces) and sparring gloves (usually sixteen ounces)

Taste
The one sense modality that I am a perceptual expert in is my sense of taste, particularly with the taste of coffee. I am able to distinguish between many types of coffee whether the coffee is Colombian, French Roast, Italian Roast, Instant Coffee, or an assortment of flavors. I think this ability to distinguish between them has to

do with my flexibility to try many types. I am a regular coffee drinker.

I am a perceptual expert using my sense modality of taste when it comes to beer. I can identify the different kinds of hops, whether they are bitter, floral, or mild flavors. Some beers have a strong flavor that most would assume have higher alcohol content when it really has a stronger hops flavor.

I am a perceptual expert when it comes to tasting and identifying different types of ice cream. I worked at *Baskin Robbins*, an ice cream shop, for two years which means that I have tasted every flavor possible.

I have always been somewhat of a spice/hot sauce connoisseur and can tell which spice or hot sauce has been put into a soup, burger, pasta; you name it, I can point it out.

Smell
I smell "people" better than most. If someone does drugs, I smell it in their sweat. I can tell if it's been one day, two days—or three days since they had a shower. I associate the smell of a person to their activities and way of life. (I am a Human Resources Manager—I think I had a tendency to smell better than most but my experience has heightened the way I use my sense of smell.)

My mother is Italian, and she always cooked DELICIOUS Italian dishes (man I miss those) but I can always pick out the smell of traditional Italian food, whether I see it there or not.

Combination
I can pick out different dance step and musical beats in Norwegian traditional dance because I have (*been*) in a Folk Norwegian dance group since I was four years old.

I can feel and see the differences in different kinds of lacrosse mesh used at the end of lacrosse sticks. This comes from playing lacrosse since I was about eleven years old.

Working at the jail I come into contact with a lot of people that are impaired by some kind of illicit drug or alcohol and I have become quite good at determining what people are impaired by. Either just by looking at them or hearing them talk.

...Then, in encountering some terrific liars and watching shows like *Lie To Me*, I became really interested in it and studied some of the facial and verbal cues people use when lying. I use my vision and hearing modalities for this and feel I've become quite good at determining when someone is lying.

Conclusion

Figure 7-11 *Left:* How closely do you look at the taxi cab driver when sitting in the back of the cab? *Middle:* Might the fact that all taxi cab *cars* look alike contribute to perceiving that all their drivers do, too? (A car can come to feel like a part of the body, Chapter 3.) *Right:* Would you notice if actor Tony De Niro from the movie *Taxi Driver* were your cab driver? Photos public domain.

There is a lot we do not yet know about how to acquire perceptual expertise. But, you now know how to make sure that to you, all cab drivers do *not* look alike.

Figure 8-5 We seem to have multiple perceptual personalities when eyes and ears disagree about what they perceive in the world. An example is the Ventriloquism Effect. Fortunately, we are rarely aware of all the conflict. Want to know where the rest of her body is? See Chapter 3, Figure 1! Original cartoon by Felice Bedford.

Multiple Perceptual Personality Disorder:
The War Within

P: "It's here".
V: "No, it's over here."
P: "It's here, I say!"
V: "You are wrong. Trust me".

A squabble between Pansel and Vretel over the location of the exit from the forest? Paula and Vera disagreeing about the location of the bus stop? Not exactly. Try the following with your eyes closed. Think about where your right hand is. You also did this in Chapter 3 when trying to mentally trace the outline of your body. How would you describe where your hand is located? Open your eyes and see if vision says it is in the same location as you described it to yourself. Note I didn't say open your eyes and see if you were correct, but rather whether vision agrees about where your hand is. That's one thing to start thinking about. Another is to note that you can determine where your hand is by looking at it or you can determine where your hand is by just feeling where it is (*propriocep-tion*). You should have demonstrated the latter to yourself when your eyes were closed. Either vision or proprioception will suffice, though often in the real world we get to use both sense modalities at the same time. You can use more than one of your senses to get informa-tion about many things, including your own body. Bang your keys on the table and you can see where they are, hear where they are, and feel where they are. But here's the key(s) observation, or question actually: What will you do if your sense modalities do not agree with one another? "P" and "V" in the dialogue above refer to the sense modalities proprioception and vision disagreeing about the location of an object. *If your senses disagree what will your mind do?*

This is what the second type of perceptual learning, that of perceptual adaptation, is about. The disagreeing sources of information about the world do not even have to come from different sense modalities. They just have to be different ways of gathering information about the same thing and if they disagree, you may get adaptation. When you opened your eyes, I'm guessing that there were no surprises and you saw your hand was right where you had felt it to be. So do you ever get into a perceptual war with yourself?

Vaudeville Returns: The Ventriloquist and The Dummy

Figure 8-1 The most famous ventriloquist of the vaudeville era, Egdar Bergen (right) with his dummy Charlie McCarthy. The act was so compelling that when Bergen's daughter, actress Candice Bergen, was a small child, she thought that Charlie was her older brother. Labels added: V=vision; A=audition. When your perceptual system registers the auditory sound in one place but the visual image of the moving lips in a different place, the conflict between sense modalities leads to the illusion that the sound is coming from the dummy's lips. Publicity photo.

Yes. In 2007, Terry Fator took top honors in CBS' *America's Got Talent*, winning a large sum of money and a show in Vegas. Just when you thought ventriloquists were an antiquated form of entertainment (Figure 8-1), they return to delight new generations of puzzled perceivers. Not only did it sound like Fator's puppet was talking, but he was able to sing while his puppet was "talking" and he had multiple puppets with different voices. If you want to know how ventriloquists make it sound like the dummy is speaking, you will find it written in several places that a ventriloquist has the ability to throw his voice. You may imagine him training for years and acquiring expertise, perhaps even like we have discussed in previous chapters. Do not right (misspelling intentional) this down, however, because it is wrong. The ventriloquist does not "throw his voice". He does not and cannot project his voice so as to originate from where the dummy is. Instead, you perceive the dummy is talking because of your own perception.

It is due to a disagreement between your eyes and your ears over the location of an object. Say you watch the stage and hear "It's great to be in Vegas". The dummy's mouth is opening and closing as you hear the words. The ventriloquist's mouth is not. Vision, therefore, localizes the speaker to be where the dummy is, let's say a little to your left at a 20°angle. However, it is the ventriloquist that is actually saying "It's great to be in Vegas", of course, not the dummy. Since there is no such thing as throwing the voice, the sound is coming from the location of the ventriloquist and that is exactly where audition detects it to be. How? Just by using the same process discussed in Chapter 2 for how any sound is localized: The difference in the time of arrival of the sound at each ear provides an accurate and precise way to determine where the sound is coming from. If the ventriloquist is straight ahead of you, for example, then the sound is equally distant from both ears; it will reach both ears at exactly the same time. If the difference in arrival time is 0, your auditory system knows that the sound is straight ahead.

A: "It's straight ahead"
V: "No it's not, it's 20° to our left"
A: It's straight ahead!
V: You are wrong. *Trust me.*

Audition is voting that the speaker is located in one place but vision disagrees and indicates that he is located in a different place (refer to the labels in Figure 8-1). When this disagreement occurs, the disagreement never reaches conscious awareness. What happens in Vegas stays in Vegas! But vision wins the silent conflict and you hear the speaker where vision says he is, not where he really is. Vision is trusted and its information overrides audition's conclusion as to where the object is—that's the *ventriloquism effect.*

Don't believe it? Try shutting your eyes and seeing if you still get the ventriloquism effect. If you don't' have a ventriloquist handy, try it at the movies. Sit closer to one set of speakers and reflect on what you perceive as you watch the movie. It will just seem like people talking; nothing out of the ordinary. But if you shut your eyes, you will suddenly hear the sound as coming from the speaker, not from where the people look to be on the screen. When vision is out of the picture, so to speak, there is no longer a disagreement between modalities. When your eyes are closed, the sound can be correctly perceived to be where it is localized by your auditory system. Open your eyes, and the sound "jumps" back to the screen. The illusion has returned because vision and audition disagree again and vision wins the sibling squabble.

The ventriloquism effect is an illusion that occurs when vision and audition disagree about the location of an object and the sound is incorrectly heard to be coming from the location of the visual stimulus. This is a well-known description yet fallacies about practicing to throw the voice remain. My students and I even tried to change an erroneous Wikipedia encyclopedia entry and someone changed it back! Ventriloquism *can* take years of practice, not at "throwing" the voice but at keeping the mouth closed while talking. If the ventriloquist moves his lips, the illusion is ruined.

The Severed Hand Illusion (and its Duller Ancestors)

Now let's try to get the sense modalities of vision and proprioception to disagree about where your own hand is. The task you did at the beginning of the chapter suggests that your senses usually agree about this. I should let you know first that there are far easier ways to create such a disagreement. They were even discovered a hundred years before the *rubber hand illusion*[1] I am about to describe but none is nearly as much fun.

Get yourself a severed hand. Any severed hand you have lying around will do. I like to time this demonstration for Halloween since the spooky atmosphere is fitting and fake rubber hands can be found cheaply. If you cannot get a hand, a severed foot works fine but a doll head (as a student of mine tried to do and then had to explain to his wary girlfriend why he had a decapitated doll in his car), not so much. Put the fake hand on the table in front of you. Place your own hand beneath the table so that you cannot see it. Every good magician has an assistant and yours must now stroke both the fake hand and your real hand at exactly the same time and rate. If done right, it should start to feel as if that hand on the table belongs to you. The fake hand begins to feel like your real hand and if you were pointing with your other hand to where you felt your right hand to be, you would point in the wrong place. For more dramatic flair, the researchers of the rubber hand illusion study took a hammer and smashed it down on top the fake hand. (I warn my students: "Please, please, do not hit your real hand. I repeat, do not hit your real hand.") Their subjects jumped in horror, much more so than people who had just been watching the demonstration. (And see our altered horror movie advertisement in Figure 8-2, which seems appropriate.)

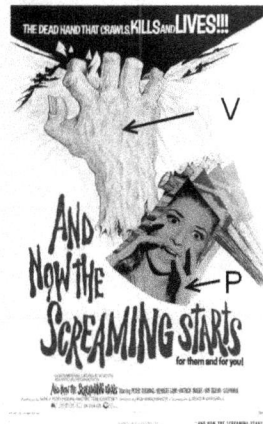

Figure 8-2 One of several horror movies starring a severed hand that takes on a murderous life of its own. We've added labels "V for vision and "P" for proprioception to this 1971 poster. If your hand feels to be in one place (P) but appears to be in a different place (V), the observer would be screaming because the severed hand feels like her own. Movie poster publicity material.

If you are a subject in the demonstration, you feel your hand being stroked in one place (more accurately, proprioception detects the hand being stroked in one place) but you see it being stroked in another (vision detects it in another). The procedure is trying to bring about such a disagreement between vision and proprioception over the location of your own hand. It is similar to the ventriloquism effect, except that it is the sense modalities of vision and proprioception rather than vision and audition that are being made to disagree about the location of an object. If successful, proprioception determines your hand to be under the table (the way you localized your hand with your eyes shut earlier) but vision localizes your hand on top of the table. If the procedure succeeds at producing this disagreement between the two modalities then, as with the ventriloquism effect, it will not reach conscious awareness. Instead, an illusion will result in which vision overrides proprioception's vote as to your hand's location. You feel your hand to be where you see it—where the fake one is on the table and not the one under the table that proprioception is insisting is your hand's true location.

The more boring, but more consistently successful, way to succeed at getting a disagreement between seeing and feeling is to look through something that changes your vision a little bit. In the 1960s, Irvin Rock and Jack Victor[2] had subjects look through the opposite of the familiar magnifying glass—a minifying lens which reduced the apparent size of whatever was seen through it. They had subjects look at small square pieces of hard plastic through the lens. While looking at a square, subjects were allowed to explore it with their hand from underneath through a cloth. The purpose of hiding the hand under the cloth was that if a subject had seen his own hand looking half its normal size, he would know he was being tricked. After simultaneously seeing and feeling a square, subjects were asked about its size. We know that people are usually pretty good at determining the size of an object just by feeling it and we actually saw that in Chapter 4 in the experiment on haptic learning. But the results were very different here when subjects were also seeing the objects at the same time.

Vision and touch were made to disagree about the size of the object because the reducing lens was giving a false reading to vision. When the conflict between sense modalities occurred, the subjects' minds resolved the bickering by trusting vision. They both drew and matched objects based on the size of the object that vision indicated, a size noticeably smaller than what the object really was. If you took away vision, subjects did pretty well at this task, too, like the haptic exploration task from Chapter 4.Yet with vision, they got it wrong. It's not as though subjects in the experiment were deliberating and saying "Ok, it looks small but it feels big; I can't report two sizes so I'll just go with vision." Subjects never noticed anything was wrong or unusual and did not even suspect they were looking through a reducing lens. The object was just perceived as one size, a size unduly influenced by vision. The disagreement between the senses produced by a vision-altering lens does not reach conscious awareness either.

To specifically get a disagreement between vision and propriocep-tion about location (rather than size), as in the rubber hand illusion, the time-honored way is to use a wedge prism. You will be hearing more than you have ever wanted to know about wedge prisms and their effects before our multi-chapter discussion on perceptual adaptation is over. A rainbow can be created by a wedge prism because it refracts different wavelengths of light to slightly different positions but a less well-known fact is that the prism also changes the position of an image overall. Viewing anything through a prism will shift its visual position in the direction of the thicker wedge edge so that it looks displaced a little in space (Figure 8-3). For instance, a traditionally used 20-diopter prism oriented with base to the right and apex to the left will shift the image by 11.3 ° to the right. That's a pretty small amount when you consider that something all the way to your side is at a 90° angle (move it any more and it's behind you), so the prism produces only about a 12.5% shift in position. That's enough though to mess up your otherwise fine coordination; try to reach for an apple really quickly while looking at the apple through the prism and you will totally bypass it, much to the amusement of your friends who enjoy slapstick comedy.

Figure 8-3 Part of a tree trunk seen through a wedge prism. Note how it is displaced with respect to the rest of the trunk. Photo by Drew Stephenson, released into the public domain by author.

Now suppose you look at your own hand through the prism. If you don't have access to a prism, submerge your hand in a clear glass of water but only look at the parts that are fully beneath the surface. Do this and you will yawn. I told you it was boring. You will just see your hand. What you will not notice, however, is that the image of your hand is shifted to the side. Your hand feels to be in one place but looks to be in another (Figure 8-4). Your two sense modalities now disagree about the location of your own hand, *un*like the very first task at the start of the chapter when they agreed. But you do not even notice, so it seems boring. You do not notice because vision overrides the differing proprioceptive signal and you perceive just one location. The location perceived is vision's location, which is indeed wrong; it's an illusion! The effect is immediate and compelling, so much so that you don't notice that anything is amiss. Once again, a perceptual war

Figure 8-4 Drawing of a hand seen through a wedge prism. "P" refers to the proprioceptive position of where the hand feels like it is. "V" refers to the visual image of the hand that was falsely displaced because of the prism. Note the similarity to Figure 8-2. The disagreement between sense modalities created with a prism is an old and reliable way of getting the illusion that the hand feels like it is located where it is seen.

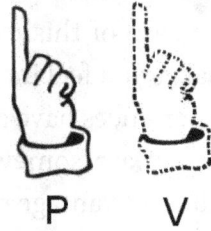

is solved behind the scenes and you remain blissfully ignorant of the internal strife.

Out of Body Experience

How far can you push this effect? Neuroscientists from Switzerland and Germany tried to create a disagreement between modalities for a person's entire body[3]. Each subject watched a live video feed of his own back being scratched. If you can imagine the situation, vision is telling you that your own self is a couple of feet in front of where you are, whereas proprioception, not to mention common sense, is indicating it is not. Can a disagreement between sense modalities that is this crazy also result in an illusion in which vision gets its way? The investigators report that there is indeed such a trend. They tested subjects by moving them while blindfolded to a new location and asked them to return to where they were originally. The subjects overshot their real location and went closer to where the false visually projected image had been. They also reported experiences such as "feeling strange".

I have mentioned a couple of times in a previous chapter the phenomenon of astral travel, in which people report that they can "leave their body". As also noted, people can perceive where they are doing the perceiving from, just as they perceive any external object. And, as with external objects, sometimes perception involving your own body is incorrect. It is incorrect if a video game has you feeling as if you are in the screen, not on the couch, or doing surgery from a distance has you feel like you have moved a thousand miles away from the controls to where the remote instruments are touching the patient. In the case of this experiment, it is incorrect if a video of your back makes you feel inches in front of where you actually are. Near-death experiences have also been reported in which people claim to feel as if they are somewhere else and can even see themselves from a different vantage point. These types of experiences start to seem less mystical when realizing that perception of our own bodies is neither fixed nor accurate. There is some older research in which people have to decide which letter, such as a "b" or "d", has just been drawn on their forehead or the back of the head[4]. This was described in an earlier chapter as well. To make a long story short, the pattern of results is that it is as if people are behind themselves facing forward to read the letters but aren't mentally turning around to read the letters. There may be a bias that your own egocenter—where you perceive yourself to be perceiving from—can more easily move forward and back while staying the same direction. Near death experiences, however, suggest it is still possible turn around to face yourself. There is little scientific work in this area, either on full-blown natural out of body experiences or on milder versions where people feel to be somewhere else through the creation a cross-modal disagreement.

One Object or Two?

But countless variations of the ventriloquism effect and visual capture of the hand with a prism *have* been conducted. Those can be used to further understand why the mind does what it does when sense modalities disagree with one another. We know that if a really big prism displacement is used, the effect of visual capture of the hand goes away: You no longer feel the hand to be where you see it and instead become aware of the disagreement in the senses. Likewise, move the puppet to be across the room from the ventriloquist and it no longer sounds like the puppet is talking. Or make the voice out of synch with the puppet's mouth opening and closing—like a badly dubbed foreign film—and no more ventriloquism effect.

Analogously, stroke the fake hand and the real hand out of synchrony and you will never get the rubber hand illusion. My students have found that the rubber hand illusion can be enhanced if it is the same color as your hand, the same size as your other hand, appears to come out of your sleeve, and is wearing your ring, rather than if its green, is a shrunken mummy hand, is visually detached from the rest of your arm, and has no personal embellishments. Similarly, for the ventriloquism effect, it was long ago reported that the more realistic, the better. For instance, people experienced a stronger illusion when the sight of a steaming tea kettle was presented with a high-pitch whistling sound in the different location than with an unexpected sound[5]. Do you think the effect would be as compelling if the ventriloquist talked to a moving orange on his lap rather than to a person-like puppet?

The findings on when visual-auditory conflicts lead to the ventriloquism effect and when they do not used to be summarized by two factors, *synchrony* and *realism*. These are both ideas that also apply to visual-proprioceptive conflicts. But an even more general way to characterize when the illusions occur is whether the perceiver infers there to be one "thing" or two out in the world. Any factor that increases the conviction that a single source is creating the signals from both sense modalities will increase the illusions whereas any factor that decreases that conviction will decrease the illusions. A green shrunken mummy hand that does not appear to come out of your own sleeve increases the chances that your mind will conclude that the hand is not yours, even if it is being stroked at the same time as your own hand. No wonder, your mind concludes, the visual position of the hand is not the same as the proprioceptive position of my hand *because it's not my hand!* That is, the visual signal from the hand and the proprioceptive signal from the hand show different locations because they are merely referring to two different hands. This is not a situation which needs to be blocked from conscious awareness. There is no need for two signals from two different objects to agree with one another. It is only when one hand—one source—has produced the differing votes for the two modalities that perceptual suppressive action is required.

No one wants to see a bad ventriloquist who moves his lips. Recall that it is the dummy's mouth opening and closing in the absence of the ventriloquist's mouth doing so that indicates that the visual image of the speaker is the dummy. If the ventriloquist moves his mouth, too, he becomes the more logical choice for the visual location of the speech than the differently-located dummy. The speaker is now just in one place according to all your senses. The differing information that you are getting from your eyes and your ears are no longer attributed to a single source, but instead to two sources: the dummy (visual) and the ventriloquist (auditory). There will be no ventriloquism effect in this situation. There has to be both a disagreement between the modalities *and* an inference that they come from a single source in order to get the illusion.

Separate the dummy and the ventriloquist or the visual hand and the proprioceptive hand by too great a distance and the illusion breaks down. This is because things that are far from one another are less likely be coming from the same object in the world. Synchrony between two modalities is also a powerfully influential factor because the odds are very low that two completely separate independent objects are doing exactly the same thing at exactly the same time. Therefore, if two sense modalities get this synchronous information, they are frequently judged to arise from a single source. If they also disagree with one another, the illusion will result. Even slight changes to the synchrony can destroy the single-source conviction and hence the illusion.

Consider more generally that what is going on behind the scenes in perception is that when all your senses are active, they cannot be permitted to confuse you. Recall those keys that you were banging on the table: How do you know that the sound of the keys, the feel of the keys, and the sight of the keys are all coming from *the keys*? Note the assumption in your perception that one set of keys is producing all three of those signals. When you bring your no-foam latte to your lips, how do you know that the cup you are feeling in your hand and the cup you see before you come from one cup and not two? I have argued in the literature[5] that this one vs. two question applies even more generally than to just comparing modalities and also ties into a long-standing philosophical and psychological issue known as *object* (or sometimes *numerical*) *identity.* When vision is disrupted—say an object goes behind a tree or even when you just blink for an instant— how do you know if the image before the disruption and the image after the disruption refer to the same object? Note also its connection to the change blindness phenomenon discussed in an earlier chapter. The problem is the same whether the two discontinuous signals come from two different times as in visual disruption, two different places, two different modalities, and even two different eyes. It is so general and frequent as to be one of the most fundamental problems in all of perception, I have argued. I have also argued that the solution is the same for all the areas.

While there are many factors that can influence the object identity decision—whether to infer different signals refer to the same object or not—only one is applicable to all of them and therefore must be at the heart of a solution: Geometry. Geometry can be applied even to time (also relevant in the next chapter), as I do in the theory I presented in the literature. The exact method of the object identity decision used by the mind of a perceiver may not matter for the purposes of the present discussion of when disagreeing modalities will lead to an illusion, so as long as it is clear there has to be such an object identity decision. One object or two? I promise then I will try to be brief. In *Object Identity Theory*, the greater the geometric transformation between the two signals, the less likely they will be judged to refer to the same object. The geometric transformations can be ordered from mild to radical: Isometric, Similarity, Affine, Projective, and Topological. These transformations come from mathematician Felix Klein's "program" of transformation geometry. Recall from discussions of perceptual expertise that when there are two stimuli, such as two views of Tony or two sheep, there is no objective right answer to the question of whether they are the same or different; the answer instead depends on what will be done with the information. The same is true here because it is such a fundamental way that perception works; the same two signals, whether from different modalities, times, eyes, and so on will sometimes be judged to refer to the same source object and sometimes will not. If there is another signal to which one can be matched from a lower less-transformed level of the mathematical hierarchy, that is the signal to whom it will match—like finding that once the ventriloquist's lips move he is now a better match for the auditory voice than the dummy. When getting signals from the different sense modalities, the area of present concern, Object Identity Theory helps understand how the decision is made as to whether or not the hand you are looking at and the hand that you are feeling are referring to the same hand. Or how the decision is made as to whether or not the voice and the moving lips come from the same speaker. And to repeat, in order for an illusion to result from disagreeing modalities, the modalities have to be believed to be providing information about the same thing.

If Pansel were talking about the location of the exit of the forest whereas Vretel was talking about the location of the witch's house, then the debate would stop cold with "Oh". There wasn't an argument after all.

Dancing Christmas Lights

Like the bigger stronger brother, does Vision always win the argument when there is one? Rock and Victor said that "vision dominates touch" a catchy phrase that has stuck in people's minds for more than half a century like a cloying commercial jingle you can't get out of your head for all eternity ("It's a pillow, it's a pet. It's a pillow-pet!" Or for my fellow geezers; "Winston tastes good like a cigarette should"). Try this: Plug in a set of blinking Christmas lights of the sort where the lights blinks on and off in random fashion. Now turn on some music. It will instantly appear as if the lights are blinking in rhythm with the beat of the music. In fact, you can dazzle your guests who will assume you have gone through a great deal of trouble to prepare a coordinated festive light and sound show (See also Chapter 13). But you have done no such thing; it is another illusion that is created by your own perceptual machinery. Can you identify the nature of the disagreement between our eternally arguing senses?

It is trickier to characterize here because the conflict is about time, rather than space, which is more difficult to picture and reason about. Ears are providing one signal: Boom Boom Boom Boom. That is, they indicate it's happening right Now...Now...Now...Now, as each beat of the music steadily unfolds. Eyes are providing a conflicting signal. Now..... Now................Now..Now, as the lights flash on and off at irregular times. The senses disagree about when the event is occurring in time, whereas the earlier conflicts were over where an object occurs in space. As with the other conflicts, the disagreement never reaches consciousness. Only one modality is assumed to be correct and only the information from that one is conveyed forward. The regular auditory rhythm wins out over the irregular visual rhythm and you experience a single unified perceptual event: hearing and seeing the lights flashing on and off together. The lights look like they are flashing in rhythm to the music even though they are not, just like the sound appears to come from the puppet's visual position in the ventriloquism effect even though it is not. Note that the object identity decision—one object or two—is present for the dancing light illusion as well. Only when the sounds and the sights are believed to come from a single source will there be a problematic conflict and the resulting resolution that we call an illusion. The sights and sounds are assumed to be part of one event rather than two, in part because the light is not making its own noise and the noise does not have its own light. If the lights also made their own sound every time they came on and off then there would be no illusion and you would correctly see the irregular flashes. If that happened, the rhythmic music sound would not capture the lights because they would be assumed to be emanating from a different source than the lights because the lights already have a sound source.

It is a wonderfully entertaining illusion even for those who do not know the explanation. When I was a teenager, I bought a radio because it came with flashing lights; despite a sale price, I paid too much for it because I thought it was something inside the radio that caused all the blue and red and yellow lights to dance with each type of music. Now there's probably an app for that. In the movie Flashdance, Kevin Bacon dances in the credits to a song that is different than the one that is played for the audience. No one notices. Imagine how chaotic it would be if conflicting information from your different senses reached conscious awareness. No one notices though because of auditory capture, your perceptual knight in shining armor.

Vision is submissive at last! Vision does not always dominate audition. All of the previous conflicts were about where something is occurring (or size, which is also a spatial extent), whereas the conflict in the dancing lights illusion is about when something is occurring. There is a good reason why vision may tend to dominate for space and audition for time. Audition is pretty good at space. The difference in the time of arrival at the two ears provides excellent information about the direction of the sound. Recall that adults have a resolution of about 1° of angle; if two sounds are separated by at least 1° of angle, adults can determine that there are two different sounds coming from two different locations. Not bad. Vision, however, is at least ten times better. Shift part of a line segment by just $1/10^{th}$ of a degree and we can tell that there are now two lines in two slightly different locations. But when it comes to time, audition is exquisite. While visual information extends in space—consider any scene you look at and it extends in all directions to fill up your whole visual field—auditory information extends in time. As you read this sentence aloud, you are getting the information over time rather than over space at one glance. The auditory system is designed to know exactly when something is occurring. Even those interaural time differences used for sound localization are a mere fraction of a millisecond, which itself is only a 1000th of a second. The visual system cannot compete with that.

The signal that is more precise for the task at hand may be the one that is trusted when there is disagreement between the senses[7]. "Sounds" very sensible, but it is not the whole story.

Things are a Little Less Clean

Up to now, one modality has won a cross-modal conflict. But sometimes, there is no complete victor. Instead, the resolution that is reached is one that lies somewhere in between the votes of each of the two modalities—a compromise, if you will. This requires some pause for thought because if one of the modalities was incorrect, and quite sensibly it would seem the less precise one, why risk incorrect perception by having a compromise reach conscious awareness? That could make perception democratic, but wrong. It is further complicated by the fact that the resolution of the conflict may depend on how you test for the illusion. Ask somebody where they hear the voice coming from and they will tell you the puppet—complete dominance of vision of audition. But ask someone to point to the direction of the sound and it is not so clear[8]; vision pulls sounds in its direction but the capture is not always complete. It has been proposed that the kind of perception you have when you are consciously aware of experiencing something is very different than the kind of not-always-aware perception you have when you act on the world through reaching, grabbing, and maneuvering within it. Some have argued there are even different major neural streams in the brain for each of these two types of perception and that ordinarily experienced visual illusions (like the Müeller-Lyer illusion seen Chapter 5) may not fool the second motor-perception system. Others, however, think those different brain pathways found are for something else and that the motor system is actually not immune from illusions when tested properly[9]. Speaking of conflicts, this one has not been resolved. But some evidence that the perception that drives motor acts can be separated from the perception that we are consciously aware of, as well as the data on what happens with cross-modal conflicts, suggest that there could be different conflict resolutions depending on how the perceiver is tested.

Returning specifically to whether the most precise modality wins a conflict, it is also not clear this is as sensible as it first seems. Perhaps the modality that is more precise turns out to be wrong. What we want is for the *correct* modality to be the victor in a perceptual dual. This is harder than it sounds. You cannot step outside yourself to see what is correct. You are stuck with perceiving the world through your eyes, ears, fingertips, skin, muscles, joints, nose, mouth, and immune system whether they end up being right or wrong. Perhaps sometimes a compromise is not so bad after all by minimizing the amount one is wrong if it turns out the precise modality is wrong. In the illusions discussed, our perceptual systems get it wrong—that's why they are called illusions. Hopefully, perception will get it right in more important situations since it can be downright dangerous to be so wrong about the world. Perhaps sometimes the precision of the modality is used as a reasonable guess in the absence of better information as to who is right. This suggests that providing a richer source of information on which modality is correct should affect the resolution. Have a difference of opinion? Bring in a third party to act as referee and ask who they think is right.

I did exactly this by adding a third modality, audition, to weigh in on a conflict between vision and proprioception about the location of a target in space[10]. The conflict was that proprioception indicates a target was straight ahead while vision indicated it was 10° to the right. In one group of subjects, auditory information was added that it is straight ahead whereas in another group of subjects, auditory information was added that it is instead 10° to the right. Since audition now agreed with proprioception in the first group, we would expect the conflict to resolve with proprioception as the winner and for the second group, we would expect vision to win now that it has audition's support. However, that did not occur. Oddly, the results were the opposite of this prediction. Discussing exactly why would take us too far afield, but there are several things to note. There is generally quite a bit of individual difference when the senses are at war; one person trusts their vision, another proprioception. In addition, information from sense modalities might only be compared one to another—no threesomes allowed. Third, "majority rules" may not always be the most accurate or implemented decision. Finally, there is evidence that it is the modality that is attended or the modality that is leading the other in the task is the one that dominates. This seems odd since which modality is leading or guiding the other depends more on the task than on which modality is correct. Overall, *it suggests that there is a lot more to who wins in this fairy tale than the jingle "vision dominates touch".*

A final clarification in this discussion of how the victor is not always so clear concerns statements like "one person *trusts* their vision" sprinkled throughout the chapter. I do not mean to imply that the person has consciously deliberated over the conflict and decided through conscious reasoning to pick vision to win (See also Chapter 12). Recall subjects are not even aware that there is a conflict. But it gets cumbersome and unintuitive to state "the perceptual system has …" Just keep in mind that much of perceptual learning operates behind the scenes out of your conscious reach, much like your digestion or breathing.

Is This Learning?

But is this learning at all? Not yet. Except for a couple of items in the previous section (the three-modality experiment and the attention/guided modality findings about which more in the next chapter), we have not been discussing the perceptual learning process of perceptual adaptation. The ventriloquism effect, dancing lights, and feeling your hand where you see it through a prism are the first reactions to a disagreement so that your conscious perceptual awareness can have as little chaos and as much accuracy as it can muster in the moment. It is still in the realm of perception, not yet learning. In a moment we will revisit the t1-experience-t2 framework for verification but first, here is a quick and dirty way to often tell which is which, perception or learning: Remove the troublemaking modality and observe what happens. If perception immediately returns to what it was without the modality then the illusion was likely perception. If perception is now different than it was before the disagreeable modality arrived on the scene then it is perceptual learning. For instance, shutting your eyes when in the presence of a movie or a ventriloquist act causes your auditory localization to immediately revert back to the speakers or the ventriloquist. Look through a prism for the first time and the visual position captures where you feel your hand to be, but quickly remove the prism or shut your eyes and the feeling of where your hand is will jump back to where it was before. Perception has not changed.

In the framework, at time 1 the stimulus is the sound of the voice directly ahead of you without any visual input: "It's great to be in Vegas". Your perception at time 1 is that the sound is directly ahead of you. The experience is the addition of the visual stimulus in a different location, the puppet with opening and closing mouth, at say 10° to your left. At time 2, perception to the same stimulus is reassessed. Your perception at time 2 to the sound of the voice directly ahead of you without any visual input is that the sound is directly ahead of you. It is identical to your perception at time 1. Remove the visual input and the ventriloquism effect goes away. But why not just define the stimulus at time 2 to include the sight of the dummy with his moving mouth and then perception will be different than at time 1? The reason is that in the framework, recall the stimulus at time 2 has to be identical to the stimulus at time 1. If the stimulus at time 2 includes the moving-mouth dummy then it has to be the stimulus at time 1 as well and then perception will there, too, be identical at both times: You perceive it 10° to your left. If the perception is different at time 1 and time 2 to identical stimuli, then it is perceptual learning whereas if the only way to have different perceptions is to have different stimuli, that's perception.

Stage direction: The curtains open to find the hand grasping a TV remote control.

Act 1, The Test at Time1
P: "It's here. Straight ahead. Life is good."

Act 2, The Experience. Enter wedge prism stage left, placed in front of the hand
P: "It's here".
V: "No, it's over here."
P: "It's here, I say!"
V: "You are wrong. Trust me". "I outrank you. We are voting that it's a few inches to the right and that's final"

Act 3, The Test at Time 2. Exit prism, stage right.
P: "Good. It's still here. Straight ahead. Whew, what a trouble maker"

This play is perception. We can turn this immediate visual capture into perceptual learning, however. Take the experience between Act 1 and Act 3 and present it for a longer period of time. Provide the same disagreement between modalities consistently and then one can start to detect a long-term change in perception that sticks around even when the conflicting modality that started it all has long since fled the scene.

Capture is the first thing your mind does when there is internal turmoil, a short-term resolution to quiet your multiple perceptual personalities (Figure 8-5, opposite start of chapter).

But continued disagreements require a more long-term resolution. Adaptation is that long-term solution. Read on to find out more.

Special Insert on a Pointing Study

I wanted to determine whether people would point with their dominant hand or use the hand closest to a target. Eighty subjects faced an array of targets equally far away decorated with colored shapes. Each subject was asked to point to a particular target (e.g., point to the yellow triangle) and then when their hand came fully to rest at their side were asked to point to the purple heart, which was the target located straight ahead. No mention was ever made about choice of hand. So in what may be the shortest perception task of all time, every subject had to point to just two targets, one could be anywhere from all the way left to the side (-90°) through all the way to the right side (+90°) and the second was always straight ahead (0°).

The majority of people used the hand closest to the target rather than their dominant hand! This was true even for right-handers (75 of the subjects). Figure 9-2 in the text shows which hand people used for each target location.

For their second pointing attempt, the majority of people pointed to that straight-ahead target with the hand they just used rather than switch to the dominant hand. For example, of people who had a target on the left side for their first point (33 usable subjects), 64% of them (21) pointed with the left hand to the straight ahead target. That's triple the number of people that would normally point with the left hand for a target located straight ahead. Overall, only 20% of subjects (15 out of 76) switched hands. They did become more likely to switch the further away the initial target, but no more likely to switch to the dominant hand (data not shown).

Finally, there was also no difference when other subjects were asked to "indicate" where a specific target was (rather than "point"). The data were the same, except for two subjects who walked over to the target rather than pointed to it….and one who used her whole hand to point, not index finger.

Overall, I found laziness trumps hand dominance! *The findings suggest precision is not important for pointing or else the dominant hand would have been selected.* This may be go very well with our follow-up study which found people were just dreadful at determining the exact location where someone else is pointing anyway. Note, though, that the experimenter stood behind each subject in the pointing experiment which, in retrospect, may have been unnatural and contributed to subjects pointing in what appeared to us to be very quickly and in a self-conscious manner. My student Tony G—the same one of the multiple Tonys fame—assisted with data collection.

9

Where Is It? Space Adaptation Changes
the Perceptual Fabric of Space and Time

I t is a fact that the things we take most for granted are the things
it would be most devastating to lose. We are trapped by space
and time yet rarely give much thought to how we perceive these.
If you get lost in a corn maze, though, or are late to the most
important job interview of your life, you may start to think about
these things. A person with vertigo feels like he is spinning and no
longer knows what is up or down or how to maintain an upright
position. Gravity may bring him to the ground where he suffers,
huddled in one spot, until the episode subsides or help arrives.
Vertigo is an extreme example of becoming disoriented and
incapacitated in the space within we usually maneuver with ease.

Lost in Space

Other people have felt the loss of spatial perception in different
ways. Bea (all names have been changed)has emailed me that she
cannot go into a supermarket alone because she will get lost. In fact,
if she loses sight of the person accompanying her, she will panic at
the thought of not finding her way out. Perhaps this feels like being
stuck in a corn maze. Jill, an intelligent professional, says that she no
longer plans trips to new cities because she gets lost. Jen, and several
others with less severe versions, report driving for miles in the wrong
direction before realizing that they must have turned the wrong way
when leaving the parking lot. Yes, most are women, but women also
are more likely to volunteer their difficulties (see also the last
question in the table 9-1!) Tim reports that buildings appear on the
wrong side of the street from where they should when he gets

home.While such *topographical disorientation* is a known result of damage to various brain regions, I have been collecting information from people of all ages worldwide who have spatial challenges but no known brain damage. There are plenty of them when you start looking and in our surveys of college students at the University of Arizona, my lab finds the following.

Table 9-1 Space, Time, and Well-being Survey

	% of Subjects Answering "True"	
Space questions	Male (967 subjects)	Female (1,464 subjects)
I consider myself to be spatially challenged	11.2%	12.8%
I have no sense of direction	9.3%	23.8%
I have trouble telling left from right	5.7%	16.2%
I get lost frequently	10.6%	23.1%
Control questions		
I listen to music at least twice a week	95.4%	97.2%
I have trouble buttoning my shirt/blouse or tying my laces	3.3%	1.6%
I am better than average at most things	85.1% *	62.2% *

Data from 2008-2012 *Question based on smaller samples of 195 men and 246 women

Many are minor spatial oddities. A common and amusing malady—unless you have it—is to confuse left and right. "Turn to the right" the gym teacher says and somehow, there is always one person in the line that turns to the left. "Your other right" has become an all to popular snide reply.Those who are mysteriously afflicted adopt little tricks to manage what comes naturally to everyone else. One of my informants reports that she keeps a scrunchy on her left (or was that right) wrist to keep track of which side is which. Another relatively specific deficit is found when a person misjudges the location of an object and overshoots it when looking with one eye on one side. This *past pointing* is not because the person is looking through a prism in a laboratory but because one of the muscles controlling one of their eyes has become recently paralyzed. This is more likely happen in older individuals, not the college students referred to in the table.

But the spatial losses can be more general, severe, and disruptive, as the examples mentioned earlier illustrate. Profound spatial loss can also occur in Alzheimer's disease. Patients find space confusing and get lost in familiar environments, later unable to even find their own home. Being spatially disoriented is not just about the inconvenience of needing a ride home. It feels terrible to wander in search of something that looks familiar but never finding it. Plus, normal day-to-day functioning becomes impossible when unable to find the kitchen from the bedroom. Mary, who does not have Alzheimer's but has other known brain damage from metastatic cancer reports to me that: "... initially I couldn't go out because if I did find the doorknob to the front door, I didn't know if I would find it coming home." Temporal disorientation is likewise problematic. Delirium is not uncommon in the intensive care unit of hospitals, believed to possibly result from sleep deprivation. It has been reported that big clocks on the walls to orient patients as to the time of day are very helpful. If you have ever woken up from a deep sleep and for a split second had no idea if it was day or night, then you know how disturbing it can be to lose your sense of where you are in time. Time is the subject of the next chapter.

For proper functioning in space, we need to know where we are located in the world at all times and where things in the world are with respect to each other. There are so many things we do "in space" that require good space perception: point out to others where an item of interest is, bring our bodies and then hands to the exact location needed to lift a glass, turn to where we hear a commotion is coming from, keep ourselves from walking into a tree branch, remain upright, walk a straight line, find our way home, scratch our own nose, and hammer a nail. These are, of course, just a small handful of our spatial behaviors.

When your perception of space is accurate, the tasks are just humdrum activities that remain in the background of what you consider to be your real life. They are taken for granted and require little of your attention. As the examples of loss demonstrate, however, to do them requires that perception of space be accurate and remain that way. Does that mean that barring some kind of damage—known or perhaps as yet unknown—space perception is the one thing at least that is unchanged by experiences in the natural world or not changeable in the lab? You are already too sophisticated to fall for that. In Chapter 2, we saw that sound location in developing infants has to change to keep up with growth of the head. For adults, too, the perception of space has to change, precisely so that it *will* remain accurate.

Space Adaptation

The stage has already been set in the last chapter. To get a change in the perception of space following experience, we begin with two sense modalities, such as vision and proprioception, your eyes and your hands, in conflict about the location of an object in space. That object can even be your own hand. If you look at your hand through a wedge prism and the hand feels to be in one place, for instance straight ahead, but looks like it is in another, such as 10° to the right, then there is a conflict (see Figures 3 and 4 in the last chapter). As we have also seen, this disagreement does not reach conscious experience. One of the senses prevails, in this case vision, and you have a single coherent perception. You ignore the proprioception information coming from your hand and you both see and feel your hand to be where you see it, even though vision is giving you the wrong information. You can think of the visual capture over the felt position of the hand as an immediate resolution to end the fight in the short term, like two fighting siblings that are immediately separated to avoid injury. But siblings return to fight again and a longer term solution may be needed to keep the peace if the disagreement continues.

The prism is a convenient way to create a continuing disagreement between modalities over the location of an object in space. It has consequently been used in the laboratory to investigate how to get the perception of space to change. The prism, though, seems so far removed from everyday life that it is easy to forget why one is using it. Putting that aside for now, what do you do in the lab to get space perception to actually change and stay changed? Anyone already well-versed in the effect known as *prism adaptation* can skip this long section to the next one entitled "Why Does Space Adaptation Occur", except be sure to look at Figures 9-1 and 9-2.

You should do this with subjects when they come into the lab

First make sure their perception of where things are around them is accurate before you go trying to mess it up and make it inaccurate. Not uncommonly, you just have people point at some objects scattered about. Good experiments use a variety of different positions. Present a target in one position, have them point to it, measure their accuracy, present a target in another position, point, measure, rinse, and repeat and repeat and repeat until you get a very good measure as to how accurate each person is. Don't wear them out too much though because you still have to repeatedly expose them to a disagreement a little later. Before getting to that part of the experiment, make sure in this part that subjects cannot see their hands when they are pointing. This may make more sense later but by keeping them from seeing their hand you are keeping the two modalities separate. You don't want vision and proprioception comparing notes and either agreeing or disagreeing just yet. This part is a test and you want to keep experience to a minimum until you provide the specific experience that you want. To keep a person from seeing his or her hand while pointing, you can either put a physical barrier between the hand and the eye, such as a cover that sits above the level of the pointing arm, or turn out the room lights and have only the targets illuminated and visible.

To be accurate at pointing without seeing one's hand, both vision and proprioception have to be accurate. If the target is at 10° to a subject's right then to accurately point 10° to the right, the subject has to see it at exactly 10°. If it is seen instead at 15°, for example, then the subject points to 15° and has a 5° overshoot. He or she also has to feel the hand to be in its correct position even if vision is accurate. Since the targets are usually out of touching range, they don't get to feel them but they do feel their own hands' position. To point at the target in the distance at a 10° angle, the hand must be brought to the 10° position. This requires accurately feeling where the hand is in space. If the hand feels like it is in a different place, for instance, 15° when it is really at 10°, then when trying to align the hand with the target, the subject will undershoot by 5°. Subjects, by the way, are sometimes wrong. They point way way off. We don't know if it's their vision or their felt hand position nor do we know how they got that way, but you usually just don't run them further in the experiment. If they are off by a little, and many people are—some people a little to the left, some a little to the right—we can work with that.

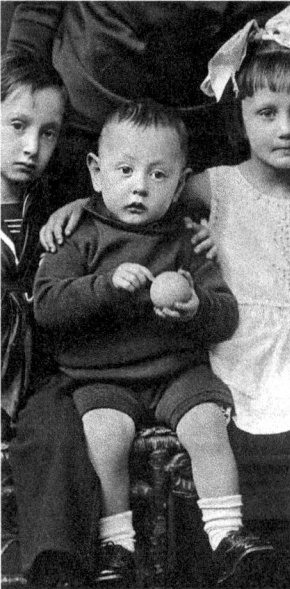

Figure 9-1 Young Sam Allalouf points to a ball, Pointing begins early in life. Photo circa 1928. Restored by Robert Bedford

Since I'm being thorough here, two more "points" while we are still on the first part of the basic experiment design. The first is a fun one. Pointing is one of those silly little things we totally take for granted. But it is actually a fascinating act that we do effortlessly. Infants already start to generate the familiar index finger extension by 4 months of age and intentionally point to objects by the time they are one year old. (See Figure 9-1.) Yet chimpanzees in the wild never point. It is a human, widespread, and likely universal behavior although dogs (but not wolves), a species evolved to communicate with humans, as well as chimps and apes in captivity, may do so as well (but never with the index finger). It is a social gesture that humans use to direct the attention of another to a particular item of interest. Infants do not point if no one else is the room to watch! See the special insert on pointing facing the start of the chapter for a study in adults and Figure 9-2 below for results. Incidentally, I gave up trying to describe the research to others in a public setting. The moment I would point to illustrate what I was doing in the experiment, everyone in the resteraunt would turn to watch. It is as hard to ignore what a person is pointing to as it is to not read printed words.

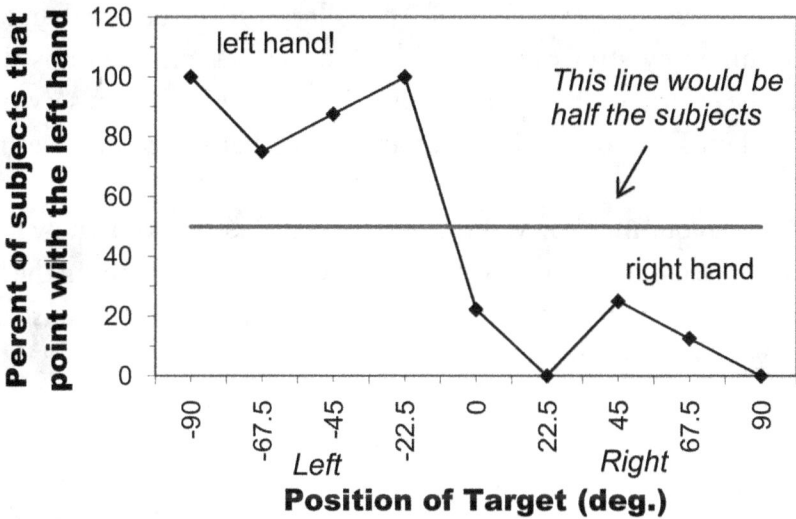

Figure 10-2 Subjects point with the closer hand, not the dominant hand. See special insert on pointing facing the start of the chapter.

The second point is to briefly consider how we can see that an object is in a particular position, such as 10° to the right. That is, how we know its position. (I hope you didn't say "we just see it"!) Recall in audition, the difference in the time of arrival of the sound at the two ears provides the cue for the angle at which the sound is coming from. In vision, this cue is fairly straight-forward when the room lights are out and the only thing visible is the target. The position of the target on the retina (the light sensitive region at the back of the eye that converts detected light to neural impulses) is determined and then it is combined with the position of the eyes. Just the position of the object on the retina would not be enough because eyes can move. The same 10° target would be 10° on the retina while looking straight ahead but would be 0° on the retina if looking directly at the target. By combining the retinal position with the eyes' position, the object's actual location can be deduced. If the head is kept perfectly still at a position straight ahead of the trunk, like it often is in a prism adaptation experiment, then those two pieces of information are all that is needed. If the lights are on then there are other tricks at our disposal to determine where the object is in space but we won't go over those. The felt position of the hand is determined by stretch receptors in the muscles and tendons that send the information to propioceptors, nerve receptors that serve the equivalent function for felt body positions that the retina does for vision.

Back to our main play, once you are satisfied that you know how accurately subjects point to targets without any experience in the lab *(pretest)*, it is now time to provide the experience that will cause the change in perception.

And then you should train them and test once more

Next, have subjects look through the prism which displaces the visual image (refer again to Figure 3 from the last chapter). An object, such as one's own hand, is felt to be in one place but is seen it in a different place (Figure 4, last chapter, and Figure 9-3, top panel). Unlike when testing subjects, now you want them to see their hands because you want to ensure that there is a clear disagreement between modalities. To get a long-term resolution to the disagreement that we can measure as adaptation, the disagreement needs to persist for a while. How long? Believe it or not, you can get a healthy effect in only 5 minutes. If you mount the prism in a pair of close-fitting goggles with no hint of the real world peaking through the sides and have people run around the world while (carefully!) engaging in normal living, it takes *longer.* It's more fun though, both

vision   proprioception

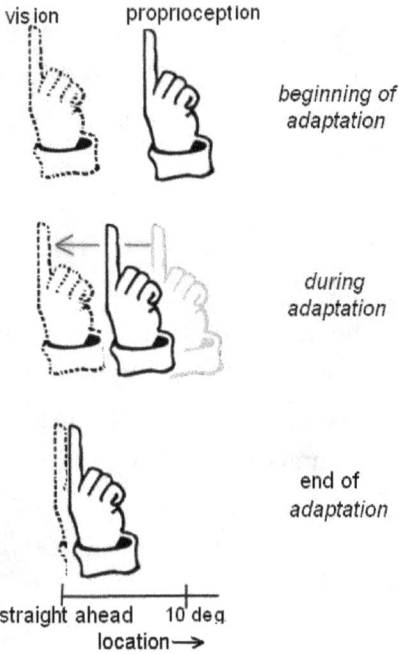

beginning of
adaptation

during
adaptation

end of
adaptation

straight ahead   10 deg
location →

Figure 9-3 Seeing the hand through a wedge prism that shifts the visual image produces a hand that feels to be in one place (10°) and looks to be in another (0°, straight ahead) During adaptation, either the felt position of the hand shifts towards the visual position or vice versa. Shown is the felt hand position shifting. By the end of adaptation, the shift in felt hand position is complete or nearly so and has realigned with vision. Note the prism in this example is shifting the visual image to the left.

for subjects and experimenters: The subject cannot catch a ball, bumps into tables, and reaches for things in all the wrong places. This happens because vision is indicating things are in one place, so when the person takes action accordingly, he or she discovers nothing is where it seems. But it takes longer because you cannot control what the discrepancy is at every moment. If instead, you have the person just watching their hand pointing from side to side as they aim for targets (yawn...), this essentially concentrates a lot of the discrepancy between vision and proprioception into a short period of time and 5 minutes is often all that you need. It also removes the conscious element of surprise because they never miss the target if their hand is always visible, an issue we will come back to.

Incidentally, there *have* been experiments where people run around the world wearing goggles packed with a distortion. The most famous may come from Ivo Kohler at Innsbruck, Austria in the 1950s and 60s[1]. He got a few people to wear goggles for prolonged periods of time that turned the image upside down. By the sixth day, at least one resumed skiing. Then again, this was Austria where people could probably ski if they themselves were upside down (and see section

"Imposters" below). I'm not sure that real skiing down a real mountain while the world has been turned upside down would pass the ethics committee these days. Many more experiments use the more sedate shift the world by 10° experience (more precisely, 11.3° caused by the popular 20 diopter wedge prism) with the subject watching his or her hand point at targets for a few minutes. Tedious, but safe and efficient.

Following the experience, repeat the same test used during the first part of the experiment. Specifically, remove the prism distortion and remove sight of the hand. Then have subjects point to the same targets as the initial test and measure their accuracy again. If there is a difference between where subjects point before and after experience with the prism then it is assumed that the experience was the most likely the cause of the difference.

Summarizng the steps of the basic paradigm

Pretest Test for accuracy at pointing to targets in space before any lab experience.

Exposure Provide experience with a discrepancy between two modalities.

Posttest Test for accuracy at pointing to the same targets in space after the experience.

What happens to perception during the prism exposure
Figure 9-3 illustrates what happens during the exposure or experience part of the expeirment. At the start of exposure, the pointing hand is localized in one place through proprioception and another through vision because of the prism. I apologize for having said this about a hundred times. Now, if adaptation is occurring then either *the visual position will start to shift to the proprioceptive position, or the proprioceptive position will start to shift to the visual*

position (or both). For instance, if the hand is felt to be straight ahead but seen at 10° to the left as in the figure, either the hand will start to feel further to the left towards vision (shown in the figure) or the hand will start to be seen further to the right towards the felt position. Either the visually perceived location changes or the felt position changes or they both change. This ends the disagreement for good. Visual capture from the last chapter serves as a temporary solution to the disagreement by suppressing the output of one modality from reaching conscious awareness when the other disagreeing modality was present. Adaptation does not suppress the output but instead changes vision or proprioception such that they are now in agreement. The change is now there whether the other modality is present or not.

If a person is still donning those stylish prism goggles, the change suits him well—it's adaptive. To appreicate this, consider a person looking through the prism for the first time and quickly reaching for a cell phone to his right. He would miss because he sees it stright ahead. In other words, he reaches for it directly in front of him, but it's not there—it is really 10° to the right. (Note at that point, his hand would be at 0° and would correctly feel like it was at 0°, but his hand would look like it was at -10° (left) because he is at looking at it, too, through the prism. This provides the disagreeable experience that enables adaptation.) Following adaptation though, the phone may now look like it is further *right* than it did when first looking through the prism if vision shifted to the right to match proprioception. That is, if adaptation is complete, and if the change occurred entirely within the visual modality (the figure shows the shift in the other modality instead), the phone actually located to the right now finally looks to be to the right as well, despite the disotrtion of the prism. The change restored behavior that is appropriate and adaptive. The subject correctly aims for it on his right now. Success!

Removing the prism, however, leads to errors! Let's see how that works in posttest.

How subjects (mis)behave during posttest

Assume the subject was accurate at pointing during the pretest before there was any exposure to the prism. At posttest, he will now point incorrectly if adaptation during exposure to the prism was successful. If the prism displacement was to the right, then he will appear to point too far to the left when the prism is removed. This is because either vision has shifted to the left or proprioception has shifted to the right. If vision shifted to the left then the targets will all be seen further to the left and the subject will point too far to the left. If proprioception shifted to the right then the targets are seen accurately but the hand is directed further to the left of the targets because the hand feels to be further to the right from where it actually is. For example, if he pointed straight at the target, his hand would feel like it was to the right of the target so he moves it further left so as to feel like he is pointing accurately. Consequently, whether the change is within vision or within proprioception or a compromise between the two, the subject will appear to point too far to the left following adaptation to a prism that displaced the image to the right. Likewise, had the prism displacement been to the left then he will appear to point too far to the right when the prism is removed.

This behavior seen when the prism is removed has been labeled the *negative aftereffect*, an unfortunate choice of term. It isn't really an *after*effect and there isn't anything interesting happening in the opposite/negative direction. Nothing changed when the prism was removed for testing. There was just one original change—adaptation—that sticks around for a while. (Indeed, work in my lab suggests it would persist forever despite claims in the literature to the contrary, provided you keep people from surreptitiously adapting back to normal. For example, keep them from touching their own body in the dark! See also Chapter 12 for comments on the longevity of various changes to perception.) The adaptive change is well suited for a displaced world and poorly suited for our ordinary world. Figure 9-4 illustrates the three parts of a prism adaptation experiment.

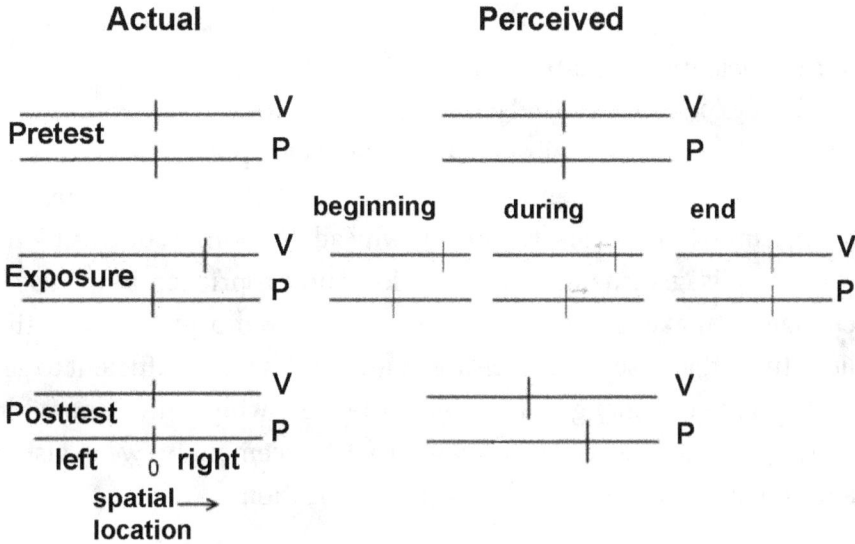

Figure 9-4 An illustration of a space adaptation experiment before (Pretest), during (Exposure), and after (Posttest) the experience with prism-displaced vision. Vertical tick marks show positions of a target through vision (V) and the felt-hand, or proprioception (P) either in reality (left column) or perception (right column). **Pretest**: A target presented straight ahead (i.e., 0°) is initially perceived accurately to be at 0° without the prism by both vision and proprioception. **Exposure**: The prism causes the visual image of the 0° target to be displaced to a position 10° to the right. At the beginning, the target is localized at 0° by proprioception and 10° by vision. During adaptation, the visual position begins to shift to the left towards proprioception and the proprioception position begins to shift to the right towards vision. By the end of adaptation, they have realigned; in this example, both meeting in the middle. **Postest**: The prism is removed and the actual target position is straight ahead. For perception, the shift left for vision and the shift right for proprioception persist and the target is incorrectly perceived a few degrees to the left by vision and a few degrees to the right by proprioception. Note the prism in this example is shifting the visual image to the *right*.

And viola! If you perform this prism adaptation experiment, you will find that the subject's perception—vision or proprioception—has changed. Our t1-experience-t2 framework can verify that and take note of the framework's similarity to the structure of the pretest–exposure–posttest prism adaptation paradigm.

Why Does Space Adaption Occur?

Brief experience caused perception of the location of an object to change. There was a disagreement between proprioception and vision and changing them restored agreement. It has been argued for a long time that this discrepancy is why adaptation occurs. But why should a disagreement between vision and proprioception cause a change to make them agree? I have presented arguments in the literature [2] that discrepancy between modalities is not sufficient to get perception to change. Consider the following three possible consequences when confronted with a disagreement between vision and touch (proprioception) concerning location.

1) *Oh-Wow.* Hey, I'll be darned. An object can be in two places at the same time. Better note that.
2) *Ho-Hum.* Vision and touch come up with two different locations because they are referring to different objects. Nothing new there.
3) *Uh-Oh.* Vision and touch localize the same object in two different places. I know that is impossible; therefore there must be something wrong with *me.*

Any of these three possibilities is logically possible. Note first that the ho-hum interpretation is the same issue of object identity discussed in the last chapter. Unless the two signals are determined to come from the same object, the difference between them is not something about which anything needs to be done. There will be no learning of any sort, perception or otherwise, just as there was no capture of one modality over the other.

If the first (oh-wow) alternative occurs, the observer will learn something new about the world. Much of the discrepant—surprising, unexpected—information that we encounter in the world is handled in this way. But the factor that precludes learning about the world and produces instead the uh-oh perceptual error interpretation is knowledge that the world cannot be the way the perceptual modalities seem to suggest. The specific knowledge is that an object cannot be in two places at the same time. Only when the disagreeing information contradicts this knowledge does the disagreement between the two modalities become a problem for perceptual mechanisms. If there is absolute certainty that an object cannot be in two places at the same time but information received from our sensory systems like vision and proprioception indicates otherwise then one deduces that there must be something wrong with the sensory systems. There must be an error somewhere in the chain of events that starts with the retina or the proprioceptors and ends with perception. The detected error leads both to the short-term resolution, capture that we saw in the last chapter, and a long-term resolution, adaptation, in which underlying changes will prevent inaccuracies for future situation. Without the one-object one-time one-place constraint on perception, there would be no adaptation, only learning about the world. Just a disagreement by itself is insufficient to produce a change in perception.

Rules To Live By

We can change the perception of space. Cool. By how much can we change the perception of space? Inspect the first diagram in Figure 9-5 and note how the prism used in the lab shifts space uniformly so that every point is shifted by the same amount. This is an approximation of what a prism actually does but it is close enough for our purposes. Straight ahead in vision corresponds to 10° in proprioception, 5° in vision corresponds to 15° in proprioception, 10° in vision to 20° in proprioception, 20° to the left in vision to 10° to the left in proprioception and so on. Everything is shifted 10° by the prism and as we have seen, perception shifts accordingly. Examples of other distortions to that one dimension of space are shown in the rest of Figure 9-5. There is no limit to what one can imagine as a change to impose upon space. But that does not mean that perception will be able to accommodate any disagreement that you

a.

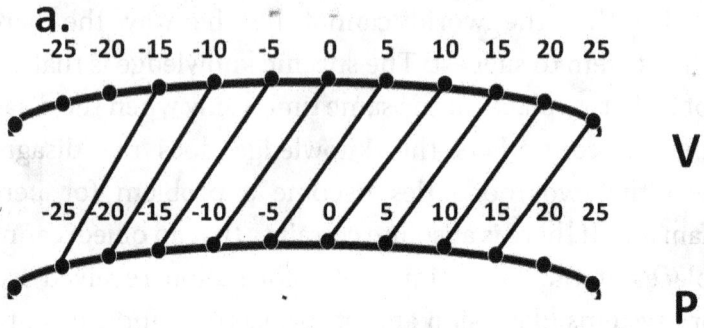

b.

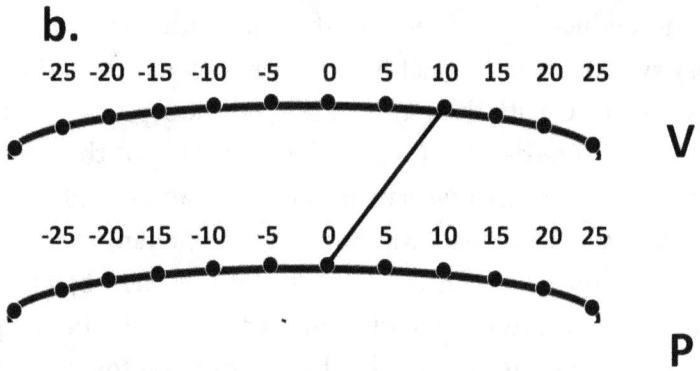

c.

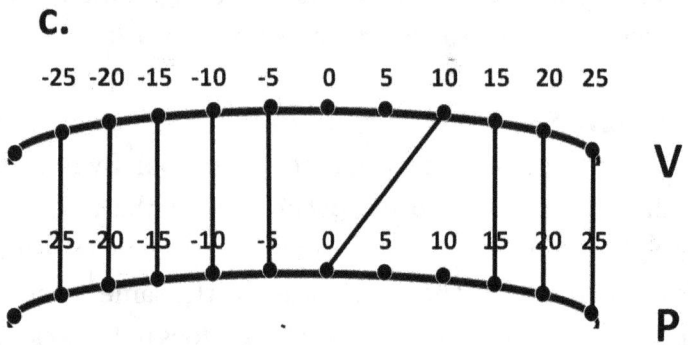

d.

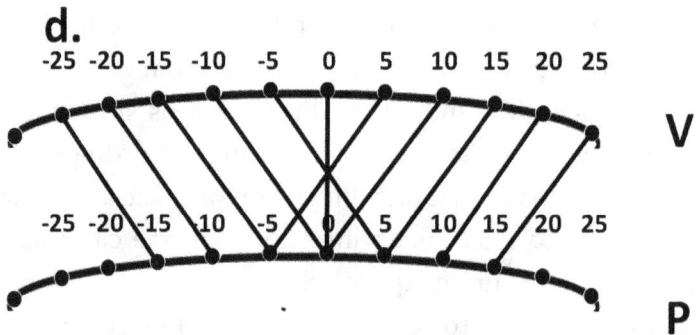

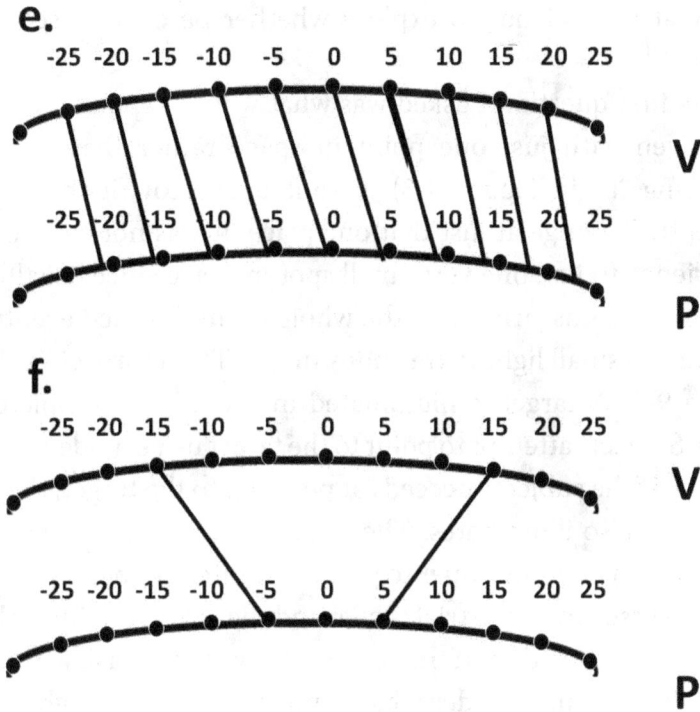

e.

-25 -20 -15 -10 -5 0 5 10 15 20 25

V

-25 -20 15 10 -5 0 5 10 15 20 25

P

f.

-25 -20 -15 -10 -5 0 5 10 15 20 25

V

-25 -20 -15 -10 -5 0 5 10 15 20 25

P

Figure 9-5 A few disagreements between vision and proprioception. There is no limit to the distortions of space one can imagine *but can perceptual adaptation handle them all?* a. Standard prism adaptation: visual image (V) is shifted with respect to proprioception (P) by the same 10° for all locations. (*Adaptation handles this easily.*) b. A 10° disagreement is specified at only 1 point in space. (*Adaptation handles easily but fills in a rigid shift at the other locations as if it received the mapping shown in "a."*) c. Similar to the last one except now the rest of space is given the normal relation in which the modalities agree. (*Adaptation cannot accommodate this and instead fits a straight line to all of space.*) d. A radical distortion in which some points that are distinct in V space are made to collapse to the identical place in P space. (*Adaptation does not master the topological destruction.*) e. The accumulated disagreement created by growth of arms and shoulders in girls from 1 to 17 years. In this region of space there are only small deviations from linearity. (*I have not tried this mapping but suspect that the average displacement, which is, 3.95 °, would be fit despite varying disagreement in this range from 3.3 to 4.25°.*) f. V space is expanded with respect to P space with only 2 V-P pairs (*Adaptation does very well and also fills in remaining locations with a uniform expansion of V with respect to P.*)

throw at it. I set out to explore whether perception sets its *own* limits[3].

The first question I asked was what would happen if experience was given with just one point in space rather than everywhere (mapping "b" in Figure 9-5). I wanted to know if there would a perceptual change at just that one place. It was not easy to restrict experience to just one very small spot in space since usually we see our whole hands, arms, and the whole room. I settled eventually on attaching a small light to the index finger. The approach is shown in Figure 9-6. A target is illuminated in space in a completely dark room. Subjects attempt to point to the target using a side-to-side arm motion. If the subject succeeds at pointing to the target, the light on the finger also illuminates. The finger light is illuminated for those moments that he continues to point accurately at the target and then after a few seconds, the trial is over and the target and finger light are extinguished together. If the subject attempts this task while looking through a prism, consider what happens. The target light turns on, say at 0°. Because of the prism, the subject sees the target at 10°. He points to 10° but nothing happens. The finger light has not illuminated because he is not pointing to the true position of the target. He was encouraged at the outset to move around until he is pointing accurately since "it's disorienting being in the dark". This is, incidentally, a small fib that subjects accept easily and they do not know that a prism is messing with their vision. It is completely dark in the room and there really aren't any clues that anything is peculiar. Subjects move their hand around and quickly stumble on the correct position, 0°, at which point the light on their finger lights up. They also are seeing the light on their finger through the prism, so it looks like it is at 10° At this instant, when a subject succeeds at illuminating the finger LED, the experimenter has succeeded in providing experience with just a single visual-proprioceptive pair. The subject feels the light on his finger to be at 0° but sees the light on his finger at 10° Thus, this approach can create a disagreement between modalities about just one single location in space. Remember, the room is completely dark so they are just seeing these few spots of light that we give them.

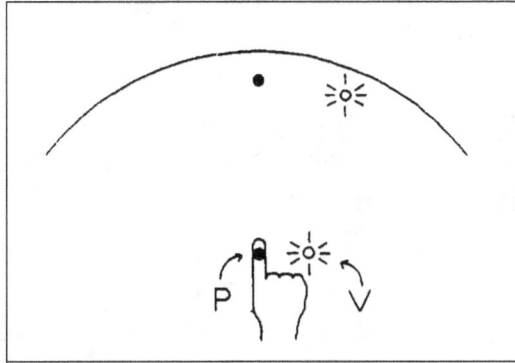

Figure 9-6 A target is straight ahead but appears to the right because of a prism. Shown is a light emitting diode (LED) on the finger that is illuminating when the subject points to the target. At this moment, the small spot of light on the hand feels to be in one place (P) but looks to be in another (V). Consequently, this paradigm allows for a disagreement between modalities to be specified at just a single location rather than an extended region of space, as occurs in conventional prism adaptation.

To test for adaptation, I used targets in a variety of different positions. When I saw the results, I thought there had been a mistake. The data looked just like I had run them through an ordinary prism adaptation experiment! The data looked like they received displacement experience for all of space and not just one spot. Subjects generalized—or really I should say their perception generalized—so that their pointing shifted from pretest to posttest at all the targets tested from 25° to their left to 20° to their right. It did *not* just shift at the place where they had the experience. For instance, experience with a visual-proprioceptive pairing of (15, 25) caused a roughly equal shift everywhere, even when tested on a target located at -25°, a position that is 40° away from the trained position.

While I let that sit there for now, consider what happens when trying to adapt to the opposite extreme. If the smallest transformation I could create to space was to impose just one little ripple at one point then the largest would be the most complicated, craziest transformation to all of space that one could imagine. There are different candidates for the honor of this title and I consequently used geometry as a guide. A transformation which destroys what is known as the topology of the space is as complicated as it gets. Destroying the topology of space would be ripping space apart or sewing it together onto itself. It would change the things we must assume would stay constant about space. The following describes part of a mapping that we tried. (The complete mapping used is shown in panel "d." in Figure 9-5.) As you read it,think about how different this is from the world in which you usually inhabit.

Suppose you look around the room and spy three things that you would like: the television remote control which is to your left, your eyeglasses which are straight ahead of you, and a banana split to your right. To obtain these items you would walk or reach to your left, pick up the remote, then move to the right for the eyeglasses, and then finally move to the right some more to grab that bannana split. However, in our new world we created, you obtain all three of these things just by staying at the one straight ahead position. You don't have to move an inch! Vision sees three different spots but proprioception feels only one. Three distinct spatial locations in vision are collapsed into one for proprioception. This transformation may be ideal for couch potatoes but is nonetheless very strange and most unlike the space that we are familiar with. What happens when perception is provided with this most peculiar disagreement between sense modalities?

I am afraid it is too radical a distortion and perception does not shift in such as way to remove the disagreement. There was never a time when subjects pointed in the identical spot for two different visual positions. What it does instead though is interesting. It shrinks visual space uniformly with respect to proprioceptive space, so either vision is shrinking or proprioception is expanding. The response is weak—weaker than if you use the same size distortions in the other direction which does not destroy the topology—but the type of response is clear. That is, perception didn't just do nothing. And now revisit the one-pair mapping experiment.With just that limited experience with a disagreement, it didn't just sit there and do nothing either. In both situations, it imposed its own change for all of space that was not what the experience provided. It seems to have a mind of its own.The changes imposed by the perceptual adaptation machinary were both linear.

Linear changes are simple and are comprised of either displacement, such that everything is shifted by the same amount, or uniform stretches or squashes, in which all of space is magnified or shrunk by the same amount. You may remember this formula for a line from grade school math class: $Y=aX+b$. For the adaptation situation, we can describe $P = aV+b$, where V is the visual position of a target and P is the proprioceptive position. Then "b" is the constant that describes the amount of displacement there is between vision and proprioception and "a" is the slope parameter that describes the amount of magnification or minification. For instance, $P = V+10$ corresponds to standard prism adaptation situation in which vision is shifted to the left by 10° and $P = 1.1V+10$ would correspond to a mapping in which proprioceptive space is additionally magnified with respect to visual space by 10%. The experiments suggested that space adaptation prefers linear changes, with rigid changes preferred more than slope changes.

We have explored a number of mappings in different ways, including using a computer monitor for the visual space with a pen and tablet that operate like a mouse for the proprioceptive space[4]. This latter equipment allows two dimensions of space to be transformed instead of one dimension—think of transforming a square shape rather than a single line of target lights. It also allows computer software to be used to create the transformation rather than being constrained by prisms and lenses. Although this procedure comes with its own disadvantages (see "Imposters" in the next section and time adaptation in the next chapter), it suggested some further preferences of perceptual adaptation machinary. As one goes past the linear transformations of rigid shifts (isometric) and uniform expansions and contractions (similarity) there continues to be a preference ordering along the same Klein geometry hierarchy mentioned when discussing object identity in the Chapter 8. Next preferred are certain kinds of non-uniform magnifications that can turn a visual square, for instance, into a proprioceptive rectangle (affine), followed by projective transformations, in which a visual square can become a proprioceptive trapezoid. Finally, topological transformations of space, such as turning a visual square into a proprioceptive circle, are preferred to non-topological transformations, such as the couch potato many-to-one mapping discussed earlier from Figure 9-5, panel "d.".

We have our answer to how much the perception of space can be changed: Changing space comes with its own rules. Linear changes easy; crazy changes, not so much.

Imposters!

I also tried something a little bit different[4]. Suppose you just ask subjects to consciously try to figure out what the spatial distortion is rather than leaving it up to perception. That's the kind of learning people are more familiar with. A problem is in front of us and we try to figure out a solution. Consider one last time a target that is straight ahead but looks to be 10° to the right because of the prism. To get the finger light to turn on, the subject's hand must be at 0° For this manipulation, I dispensed with the prism entirely. A target was presented at 10° and the subject's hand had to be at 0° to illuminate the finger light. Software created this offset rather than the prism. Note how both here and in the prism paradigm the target is seen at 10° but must be pointed to at 0° to perform the task successfully and illuminate the finger light. Instead of telling people they may have trouble initially because it is disorienting being in the dark, they were told that they should try to figure out where to place their arm for each target. The mapping that subjects were given for each target position is shown in panel "c." of Figure 9-5. During posttest, they were instructed to reproduce what they had figured out as each target was presented.

Let's look at how subjects do compared to subjects that get the same mapping, but with the prism. Consider first the results of the prism adaptation group. The mapping specified that at one position, 0°, vision and proprioception disagree by 10° but at all other positions, there is no distortion at all. Seems like a pretty easy rule to learn—except that perception did not learn it. Figure 9-7 shows what actually did happen. Subjects' pointing shifted at all positions even though the mapping specified that only one position should change.

Figure 9-7 These two results illustrate that perceptual adaptation has different rules than other types of learning processes that do not change perception. For both groups of subjects, I specified that visual space was displaced from proprioceptive space by 10°at a single central location while the rest of space was given the usual normal relationship. The Intentional learners (bottom) were able to learn to isolate one region of space, but adaptation subjects (top) instead pointed everywhere by the

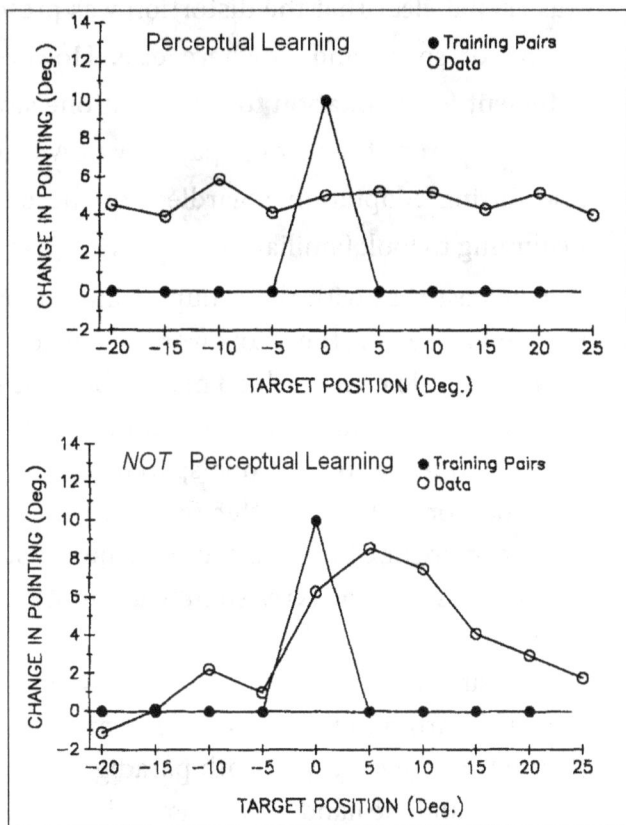

same amount. Filled circles show the mapping that was specified and open circles show what subjects actually did (shown is change in pointing from pretest to posttest).

The exact amount that it shifted appears to be an average of what was experienced at all locations weighted by the number of trials at each location. Subjects had received 50 exposure trials at the 0° target and 50 trials total for all the other positions combined. Recall in the first experiment discussed that made use of the finger light paradigm, a distortion at one spot was also imposed by us except then, no information was given about the remaining locations. In that experiment, adaptation occurred as if it were assumed that all locations must have been shifted. In this experiment, the mapping was crystal clear that the distortion was present only for one small region of space and nowhere else. However, this was still not sufficient for adaptation to single out one small region and treat it separately from the rest of space. By now though, fitting a straight line during adaptation regardless of the distortion imposed was beginning to look familiar.

Contrast that with the group of intentional learners. As can be seen in Figure 9- 7, these folk were able to learn the simple rule. The data are a little messy, which may be because subjects are doing the task in an unnaturally dark environment. Nonetheless, it is clear that they could learn to treat one spatial region differently. This world learning, for lack of a better term, is playing by different rules. It appears more flexible to new things whereas adaptation prefers linear changes to space and is resistant to attempts to take space apart point by point.

The similarities and differences between the conditions that lead to world learning and perceptual learning can be seen by comparing Figure 9-8 to Figure 9-6. Both paradigms show the visible target in one place and the hand in another. Indeed watching subjects pointing during exposure would not reveal which group was which. Why then isn't this also adaptation? After all, subjects also improve at initial pointing errors and even end up being more adaptive in a laymen's usage of the term "adapt" because the rule was learned better. The critical difference is also seen in the figures. Unlike the perceptual

Figure 9-8 This paradigm used for intentional learners is similar to the perceptual learning paradigm (Figure 9-6) because the light appears to the right but a subject's hand must be straight ahead to illuminate the finger LED. However, note there is no discrepancy between the 2 modalities. Vision (V) and proprioception (P) instead agree about the finger's location.

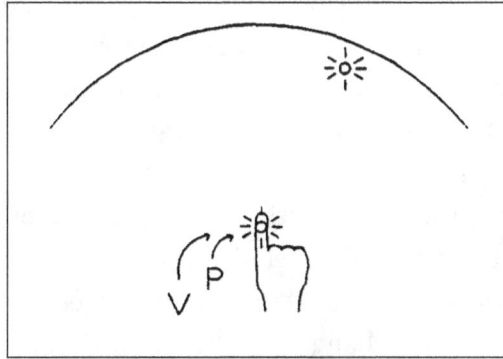

learning paradigm, in the world learning paradigm, the subject both sees and feels his hand *in exactly the same place* even if he must learn to move his hand to a different place than the target. He must learn something new to succeed but there is no disagreement between where he sees and feels the same object. That discrepancy is a necessary condition for getting perception to change. Without it, there can be no genuine adaptation. Without the disagreement between modalities over the same object, there is no internal perceptual error that needs to be corrected. One object is not detected in two places at the same time so all is well with the internal works. Other kinds of learning can take place, but not perceptual learning.

Fortunately, it is easy to realize with this intentional world learning experiment that the necessary type of discrepancy to get perception to change is absent from training. Other impostors to adaptation are not so easy to spot. In a regular prism adaptation experiment, if subjects become aware that their vision is being distorted then they may try intentionally to find a solution. A subject who realizes that he has pointed too far to the right may quite sensibly correct himself and remember to intentionally point further left next time. Just like a bowler with a gutter ball on the right side will try to aim further left next time, so too, will someone who overshoots his target aiming to the right. Early researchers of prism adaptation realized this was a problem and discussed such *conscious correction*. It is a problem because if the subject persists with his leftward correction even after the prism is removed, he will point too far to the left during posttest. This will make it appear as if he has a "negative aftereffect" from successful adaptation. However intentionally pointing further left is not adaptation. For adaptation, he has to really feel like he is pointing *at* the target yet be pointing wrong because either where he sees the target or where he feels his hand has changed. If the negative aftereffect occurs because he is trying on purpose to place his hand somewhere else, the results of an experiment may be mistaken for adaptation. As we have seen, these other systems, such as world learning, have different rules. We want to understand how perception changes, not something else.

There are several approaches for avoiding faux adaptation. Designing a procedure in which subjects never make any pointing errors and never even know there is a distortion is effective. This removes all reasons a subject would have to intentionally alter their behavior. Many experiments, however, use rapid initial reaching movements in a fully illuminated environment while adapting to the prism. Subjects given these procedures do make pointing errors and they do notice these errors. If the hand is initially out of sight and then brought forward as it aims quickly for an object (*terminal pointing*), the subject will indeed miss as he looks through a prism. Despite the old opaque jargon of "terminal pointing", this is a very natural way that we reach for objects in the world. Not surprisingly, there are lots of tools at a perceiver's disposal in the event reaching misses the object. Successfully obtaining a desired object is too important to leave it to one system. Nonetheless, if we don't isolate the adaptation system, we can end up studying something else.

Another approach is to let people notice the errors and the prism distortion, but then let them know explicitly when the prism is removed. This way, if they learned a strategy for dealing with the prism they can stop using it when they know it is no longer needed. You can even give them an opportunity to show you what they learned consciously later, so they don't feel cheated and they get to show off. However, subjects may have gotten used to correcting their aim and now it is hard to remind themselves to stop compensating for a prism that no longer exists. Habits are not adaptation either. Moreover, when subjects become aware that an external device is the cause of an erroneous visual signal, does this prevent the real adaptation process itself? After all, they now have a different reason for a visual-proprioceptive disagreement. The disagreement can—quite rightfully—be blamed on the experimenter rather than perception. While we would hope that adaptation is protected from such dangerous conscious ideas, I am not so sure that it is. My own experiments in which subjects do not become aware of experimenter-created distortions lead to nearly complete adaptive changes while many others in the literature reach a maximum of only 50% of the imposed distortions. There are, however, other aspects of the finger light procedure that may encourage such complete adaptation.

A third approach to ensuring an observed pointing change is not due to an imposter is to directly test for a change in vision or in felt hand position. In genuine adaptation, the negative aftereffect observed to occur when the prism is removed occurs either because visually perceived location or felt perceived location has been shifted. Directly assessing vision or proprioception bypasses the fake reasons for a pointing error. This is the perfect time for a brief digression that should be familiar because of the last chapter: If proprioception and vision disagree, who has to change? Proprioception or vision?

Charlie Harris was one of the gods of the adaptation boom of the 1960s. He devised a battery of tests to assess which modality is responsible for the negative aftereffect when adaptation does occur[5]. In one task, subjects who are trained with the prism while pointing with the right hand are to be tested with the left hand. If the left hand shows the same error as the right hand when pointing to targets then this implicates vision. A change in where the target is seen means that anything you use to point to it, whether it be right hand, left hand, or a foot will point in the wrong place because the target is seen in the wrong place. A different test in which the left hand points to the right hand in the dark rather than to a target in space will show no errors if the change was in vision because there is nothing to look at. On the other hand (so to speak...), if the change was in the felt arm position, then the left hand will point to an incorrect location because the right hand feels to be in a different place than it actually is but the left does not.

He found, incidentally, that side-to-side motion while pointing to targets (*concurrent pointing*) while looking through the prism, the kind in which subjects do not become aware of any errors, led to just about exclusive changes to the felt position sense of the hand seen through the prism. Terminal pointing described earlier (the in-and-out movement, where subjects can notice errors) tended to add a change to vision as well, more specifically because of a change in the felt position of where the eyes are in the head. Recall that position of the eyes in the head is one of the factors that goes into determining the location of an object using vision. Other work has added details to whether vision or proprioception wins the war. The factors are the same as those discussed for capture in the last chapter. Precision is one of those, so often vision wins and proprioception changes when these two modalities are in conflict over location. Vision is more precise about where an object is in space than is feeling or pointing to it with the hand. But inexplicably, attention/guidance is another one. The example that Gordon Redding, another prism guru, likes to use is to consider checking the time by looking at the watch on your wrist. (We know from our own surveys, shown in the next chapter, that 20% of men and 16% of women still wear wristwatches...). Since the hand is initially unseen, it is proprioception that is guiding the motion (or is "attended to" in the older versions of his theory) to perform the watch behavior. The guided modality dominates. He has evidence that that is why, for instance, terminal pointing tends to cause vision to change[6].

Returning to the issue "at hand", tests that isolate vision or proprioception as the cause of the pointing error can help rule out imposters. Since tasks such as pointing with the other hand are different than what was learned during exposure to the prism, neither habits nor conscious correction will cause them to change.

One final imposter will be mentioned. Charlie Harris introduced this in a paper *Beware the Straight Ahead Shift*. We are an enthusiastic bunch, aren't we? I met him long after he had stopped working on adaptation and was at Bell Labs, a think tank that paid people pretty much to do just that—think. He gave me a tip, which I pass on to you, though I have still lost count of tip numbers. He suggested keeping a list in the top desk drawer—you will have to translate that into the modern storage equivalent—of your top ten interests. He showed me his list, although I do not remember what was on it. But I do remember him saying that such lists are interesting because they keep changing even though you don't think they will. This has proved a useful tip for my students as well as me. I give it to my students often and it is rather informative. Those who go on to do well are often those who embraced this task; what could be more fun than getting to write down things that make your intellectual heart beat faster? But others hate it or are puzzled by it or write down the same thing ten times. Rarely do those people continue with a life of intellectual discovery. And he was also right: Your top ten lists of interests really do change over time. I will spare you mine.

The *straight ahead shift* is an artifact that results from seeing things that are lopsided in space. Things that you look at right now in your world are pretty much evenly distributed with some things on your left and others on your right. If someone were to look at things that were primarily on one side for a few minutes then what would happen is his or her concept of straight ahead, that is, the midpoint that bisects left and right, would shift further to the side that had more stuff in it. Uh, oh. Here is another shift that does not result from a discrepancy between modalities because you don't even need two modalities, let alone a prism, to get the straight ahead shift. Yet this can be a problem for prism studies because the prism shifts the images of things to one side. If the test parts of an expeirment involve only pointing at targets in different spatial positions, the straight ahead shift should not taint the results of the experiment. But some studies test subjects by having them point straight ahead (*proprioceptive straight ahead*) or say "stop" when a visual target moves into a straight ahead position (*visual straight ahead*). They do this to try to assess whether an adaptive change is due to proprioception or vision. However, straight ahead shift is an imposter that will cause errors to be made on these tasks in posttest and therefore be mistaken for adaptation. The best approaches are to avoid straight ahead test procedures, or to not rely on them exclusively, or select targets during exposure that look symmetrical while the prism is being worn.

All of the strategies mentioned here in the last few pages have been used at one time or another in space adaptation studies to ensure that a negative aftereffect is really due to adaptation. Modern experiments forget though and it is important to ensure a procedure produces genuine adaptation before incorporating a new finding into what one thinks has been discovered about adaptation.

Treatment for a Spatial Disorder

About now you are probably wondering if there is any practical use for adaptation. Actually, you have probably been wondering that for quite some time. After all, the other perceptual learning process, perceptual expertise, provides experiences of new things about the world never before detected and skills useful for a profession as well as for enjoyment.

Hemineglect is an intriguing disorder of space perception mentioned in the very first chapter which was about perception taking a wrong turn. A patient with hemineglect ignores one side of space. He may not notice people or objects from that side and behaves accordingly, such as only eating from half his plate. The ignored side is usually the left and results from damage to the right side of the brain. Equally intriguing was the idea of researcher Yves Rossetti and colleagues to try prism adaptation as a treatment for hemineglect. When Rossetti visited my lab years before, I remember him like a sponge, eager to absorb everything he could learn about prism adaptation. I am glad he did.

They had patients look through a prism that shifted the image to the right while pointing to visual targets[8]. They found a big improvement in a measure that hemineglect patients normally do poorly on, that of determining what is straight ahead. Hemineglect patients initially point way too far to the right when asked to point straight ahead. This is certainly understandable given their entire world is to the right. Following prism adaptation, there was an aftereffect such that they now pointed further to the left. The effects seemed to last longer than existing treatments for hemineglect, such as instructing a person to gaze to the left. It therefore led to an explosion of work on prism adaptation and hemineglect, largely with the aim of using it as rehabilitative treatment. There have been differing opinions on the outcome, some suggesting prism adaptation is the most effective treatment, others suggesting it only helps some people or needs to be used in conjunction with other therapies.

Surprisingly, I do not know of any study that assesses whether it is indeed prism adaptation that helps hemineglect patients. That is, it could be that adaptation is just an innocent bystander and it is really one of the imposters that is therapeutic. There may be something about a prism adaption procedure that gets people to act within the left half of space which could be used equally effectively without obtaining adaptation at all. For instance, it has been noted that the prism adaptation procedure forces the patients to intentionally move their arm to the left in order to compensate for the right image displacement effects of the prism. This sounds a lot like the conscious correction imposter discussed earlier. Suppose one of the other procedures for adaptation discussed earlier was used instead in which conscious correction never comes into play. Have subjects point to targets with a side-to-side motion such that the hand is never out of view of the prism (concurrent pointing). The procedure routinely used for hemineglect studies is instead terminal pointing in which subjects notice aiming errors. Recall, that is the one in which a subject's hand is initially out view and brought forward to point to a target.

If it is the case that genuine adaptation is what treats hemineglect, then patients should see benefits with either procedure. If instead it was the conscious leftward correction of pointing movements, then only the terminal pointing adaptation procedure would be successful. That would be a very informative test that has not to my knowledge been conducted. Another likely candidate for an imposter would be Harris' straight ahead shift, also discussed earlier. Could it be that what helps is just looking at an array of targets that are further to the right than usual? If so, just presenting asymmetrical targets without a prism, or even better, trying prism adaptation again but ensuring the targets remain centered, would unravel that confound. Perhaps consistent with the hypothesis that what really shifts is the *concept* of midpoint, not perception, is Rossetti's additional finding that the midpoint of numbers shifts as well. It sure would be an ironic, but fascinating, treatment for someone who sees things on the right to show them things that are even further to the right.

Can you pretend to be a patient? That's how I started this entire perceptual journey which I am taking you with me on. If you had hemineglect, would you care if it really was genuine adaptation or an imposter of intentionally moving your hand to the left or seeing targets to the right? Not one bit. The procedure helps, so why look more closely and risk ruining it? But, of course, if you are a science geek that mounts prisms in goggles, you care a lot.

There is an application of prism adaptation research that I think is very important. I have devoted a whole chapter to it (Chapter 11) and you will find it following the next chapter on the time counterpart of space (Chapter 10).

Why on Earth (or in Space) Does Adaptation Occur Again?

The everyday value of adaptation can also be addressed by digging deeper into why adaptation occurs. For perceptual expertise also, we revisited the purpose of a mechanism that enabled perception to improve with practice. To retrace what we uncovered so far about why adaptation occurs, five factors are necessary to get adaptation. 1) The same perceptual properties, such as space, can be apprehended in more than way, such as through both vision and touch. 2) The signal received from each system can be compared to one another. 3) The decision as to whether or not the two signals originate from the same source is a key part of perception. 4) If two signals judged to refer to the same source disagree, there can be different outcomes, not all of which change perception. 5) If the disagreement contradicts prior irrefutable knowledge that this is an impossible state of the world then the disagreement is attributed to an internal error and one of the involved perceptual systems is adjusted to fix the error. It is this perceptual change referred to in the last item that we call adaptation. As seen, the set of steps culminating in the fifth and last item help explain why there is a change in perception rather than a change in knowledge about the world, or no change at all. For space, the knowledge is that one object cannot be in more than one place at one time. If the signals from vision and proprioception indicate something to the contrary, that one object is in more than one place at one time, then the disagreement between them is attributed to an internal error in vision or proprioception. Visual space or proprioceptive space is tweaked to fix the machinery.

That is what we know so far. Now to progress deeper, why in turn do we have such a strong constraint on perception, from at least early infancy, that one object cannot be in more than one place at one time? Wouldn't it be discovered soon enough when interacting with the world? And more so, why have a mechanism waiting to fix the problem that two modalities indicate that an object seems to be in two different places at the same time if that situation is physically impossible? Surely adaptation cannot be a mechanism just lying in wait for an artificial disagreement created by a white lab coat and a pair of prism goggles. There must be an important real-world situation that produces a disagreement between modalities over the location of an object. Without the one-object one-place one-time constraint, we would draw the wrong conclusion about what the modalities were telling us with their disagreement and without the machinery to adapt, we would be unable to fix the broken perceptual system.

Growth is the answer that Richard Held (the same 50 years of research and still going strong Richard Held) offered a long time ago[9]. We saw in Chapter 2 that physical growth of the body shows that perception has to change. If a child's head grew but adaptation was non-existent then errors in sound localization would go un-fixed. Interaural timing differences of the sound at the two ears that worked for a smaller-sized head would still be used for calculating the angle of an object. This would lead to errors when used with a larger size head. Growth seems like a sensible explanation for the existence of adaptation. The head is not the only thing that grows that might be expected to affect the perception of the location of objects in space. The shoulders get further apart, which should affect proprioception, and the eyes get further apart, which may affect vision. If modalities grow at different rates then this would be a naturally occurring, important, and common situation that would cause a disagreement between modalities. An object could appear to be in two places at the same time by the different modalities while we are growing!

This explanation that adaptation is "for" growth has sat untested for half a century; it makes so much sense. If the mechanism responsible for such easily obtained adaptation to prism-displaced vision in the lab really serves the function of accommodating growth-caused spatial distortions in the real world then there should be a close match between these two problems. I therefore set out to determine precisely what the type of adaptation would be needed to maintain accurate pointing to objects in space when faced with having to grow. Would it match the ease with which linear changes are accommodated in laboratory adaptation studies? Are uniform displacements where all of space is pulled in the same direction by the same amount also predominant for growth the way the are in the lab with a prism? A casual analysis seems promising; as the shoulders get further apart, the right arm, for instance, is displaced to the right. Before adaptation to the new body size, the hand would feel like it is further to the left than it actually is. This would cause a person to point too far to the right of a visual target. To restore accuracy, the felt arm position would have to shift to the right, seemingly not unlike what happens in prism adaption. To look more closely, I derived the changes to pointing that would occur with growth and therefore, the type of adaptation that would be needed to accommodate the growth and maintain accurate pointing anyway.

Math modeling of growth

I found that the adaptation that would be needed for growth is dependent both on the width of the shoulder and the length of the arm. In particular, it is the ratio of half shoulder width (mid back to tip of shoulder) to arm length both before growth (R1 in the equations below) and after growth (R2) that determines the plasticity that is required by growth. The result of the math modeling is shown in Equation (1) and the derivation of the equation is shown in the appendix. It is an ugly equation which already spells trouble.

In Equation (1), the quantity $\alpha1$ is related to the angle of the target and $\alpha2$ is related to the angle at which the subject would point following growth and before adaptation. (Specifically, $\alpha1'$ is the angle of the target and $\alpha1 = 90 - \alpha1'$; while $\alpha2'$ is the response angle and $\alpha2 = 90 - \alpha2'$). As noted, R1 is the ratio of half shoulder width to arm length before growth and R2 the ratio after growth.

$$\alpha2 = \tan^{-1}\left[\frac{\sin \alpha1\left(\sqrt{1 - R1^2\sin^2\alpha1}\right) + R1\cos\alpha1}{R2 - R1 + \cos \alpha1\left(\sqrt{1 - R1^2\sin^2\alpha1}\right) + R1\cos\alpha1}\right] \quad (1)$$

The result is s a little prettier when kept as two separate functions

(Equations 2 and 3):

$$\gamma' = \alpha1' - \sin^{-1}(R1 \cos \alpha1') \tag{2}$$

$$\alpha2' = 90 - \tan^{-1}\left(\frac{\cos \gamma'}{R2 + \sin \gamma'}\right) \tag{3}$$

where γ' is the shoulder angle that is required for accurate pointing to the target. That is, before growth, present a target at angle $\alpha1'$, and then γ' is the angle of the shoulder required to point to it accurately (equation 2); but if this shoulder angle for a smaller size body is erroneously used after growth, then $\alpha2'$ is the angle at which the arm will wind up pointing in attempt to point to the target (equation 3).

Have you ever given much thought to how long your arms are? Or how wide your shoulders are? Fortunately there are many people who have. Clothing manufactures for one! As a result, there exist tables of data on the size of just about every body part there is. Growth studies have also been undertaken to establish norms for children so as to identify individuals with abnormal growth. Data also exist for cross-cultural comparisons and cross-era comparisons. If it can be easily measured, someone has measured it. I consequently looked through tables of body size norms at different ages for the body parts[10] relevant for the equation and then took the ratio. The ratio as a function of age is shown in Figure 9-9.

An adult has a ratio of (half) shoulder width to arm length of about .25, so half shoulder width is one fourth of arm length. The whole width of your shoulders is about half that of your arm length. The ratio in women is very slightly larger than in men. That is, a man has proportionally longer arms than a women compared to shoulder width. And here you thought men had broad shoulders when really the long arms are more notable... If you inspect the ratio as a function of age in the figure, note that the shoulder width to arm

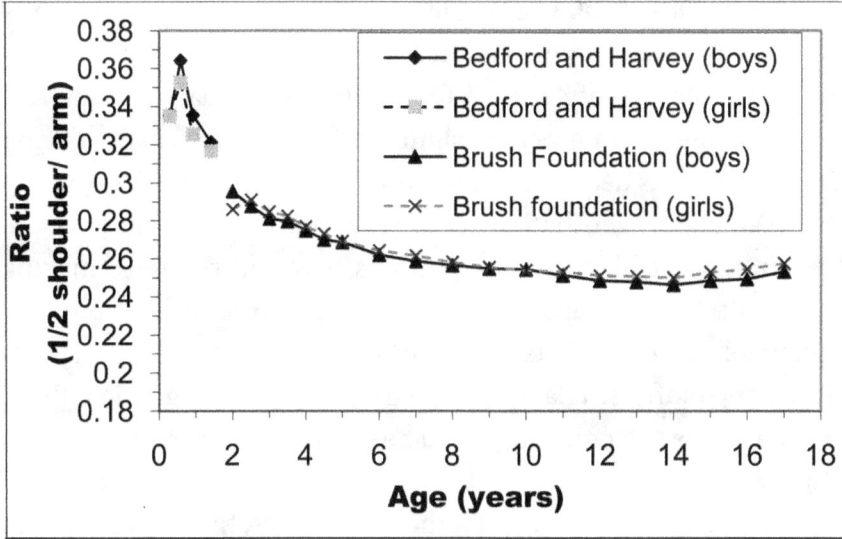

Figure 9-9 Babies have short arms. The proportion of shoulder width to arm length drops during growth. By adulthood, the arms are about twice as long as the shoulders are wide. Interestingly, the proportion is the about the same at age 8 as it as at age 17. This implies that even without the ability to adapt to growth after age 8, pointing to objects in space would be accurate as an adult. (And note from the graph: do baby girls have proportionally longer arms than boys, a relation between genders that reverses sometime between age 2 and 2.5 years?)

length ratio at age 8 is the same as that of an adult. Now take a look at equation (1). If the ratio before growth and the ratio after growth are identical—R2=R1—then everything cancels and the equation reduces to α2 = α1. That is, since the ratio of shoulder width to arm length remains the same, no plasticity at all is required! Pointing to a target in space will be just as accurate as an adult without any adaptation despite the large amount of physical growth that occurs from age 8 to adulthood. Interestingly, developmental psychologist Ferrel and colleagues say: "The turning point character of age 8 in development is, once again, emphasized..." [11]. They also single out age 8 as special in pointing when they find an improvement in motor capacity, a decrease in intra-individual variability, and an increase in sensory-motor integration. Adult-like pointing may coincide with adult-like body shape: Our trunk is a rectangle which is twice as long as it is wide both as a nearly grown adult and at age 8.

Between the ages of 8 and 17, the ratio does fluctuate, but slightly. It keeps on decreasing before climbing back up, especially during adolescence. This decrease in ratio reflects that the arms are getting proportionally longer than they have ever been or ever will be again. These non adult-like body proportions may contribute to making adolescents look peculiar. It may also explain a general physical clumsiness in adolescents and why they report that their sports performance suffers. That's because the small changing ratio implies that small errors in pointing would be produced without adaptation until they reached adult proportions. Likewise from age 2 to age 8, there are changes in the ratio. At these early ages, the ratio decreases from a high of .295 in males and .285 in females; these changes, too, are quite small, especially compared to the amount of absolute growth involved. The decrease in the ratio would cause a small pointing error to the *left* with growth (not to the right as the initial superficial analysis suggested). For instance, for a straight ahead target, the accumulated error for girls from age 2 to 17 if there were no adaption would be only 1.75° to the left for a target straight ahead and 1.41° for a target located at 25° to the right, much smaller than a 10° prism.

Before the age of 2 years, our own data (a collaboration with former student Erin Harvey, now on the faculty of Ophthalmology at the University of Arizona and the same Erin mentioned in the story in Chapter 5) finds that bigger adaptation changes would be needed to maintain accurate pointing—but it's not clear how accurate infants and young children start out at pointing at 11 or 12 months of age. Some other work I have done suggests that even adults' comprehension of where others are pointing is so bad that huge errors in pointing can occur without undermining its social communication purpose. If the small changes between age 2 and age 8, or the larger changes from infancy to age 8, are adapted to then the changes required are not the same for all the different target positions. That is, it is not a rigid shift of all positions like prism adaptation, nor is it a uniform magnification or minification. Instead of being linear, the error as a function of target position is described by the complicated ugly function shown in equation (1). Over a small central range of space though, like the one that tends to get used in experiments, the deviations from linearity are not too noticeable. The disagreement between vision and proprioception that would be produced by growth of shoulders and arms for girls from age 11 months to age 17 years is shown in figure 9-5, mapping "e.".

In sum, there was, rather unexpectedly, a poor fit between the demands of growth on pointing to distant targets (non-existent to minimal, even when allowed to accumulate, and non-linear when present) on the one hand and adaptation to space in the lab (large, easy to obtain, linear especially uniform displacement) on the other. If adaptation is not for growth, then what is it for?

Modeling and thinking about just moving around
I found a much better fit through just simple body movement[12]. If you rotate the shoulder a little bit backwards, let's say by 10°, but your mind has not yet caught up to this movement then this would produce precisely the uniform displacement produced by a wedge prism. In Chapter 3, body schema was the topic of discussion. Adaptation seen in the lab may be more about maintaining accurate body schema in the face of ordinary (which are linear!) movements than it is about growth. In general, our bodies cannot remain at peak performance without continual calibration. If even a very accurate time piece will fall out of synch with the atomic clock in Boulder Colorado, unless a signal is passed between them, I am not sure we can expect our bodies to do any better. Small errors in accuracy would occur over time that would accumulate if not corrected.

The answer to our "why" question then is: The only way to keep perceptual systems accurate *on a day-to-day basis despite motion of the body* is through the internal error detection and correction provided by adaptation. For space perception, this means maintaining accurate felt position of all body parts at all times, including eye position, a challenge made even harder by the ability to rapidly move ourselves around in space. Sure, that purpose of adaptation is not flashy (but wait: see Chapter 11!). Be grateful though, because without keeping your perceptual systems accurate and happy, they would feed you all the wrong information about the world!

If space adaptation is more for the day-to-day maintenance of accurate body schema when moving the body around throughout life rather than growth of the body during childhood development, then what is the relation between growth and adaptation? There are a number of different possibilities. One possibility is that growth may have its own specialized adaptation mechanisms. Consistent is that there are some animals that can only adapt when young. Another possibility is that it's just pointing that need not be accurate and other growth problems would be a better fit to our linear adaptation mechanism. This may be the case with audition.

Modelling growth and hearing

I also derived the adaption that would be needed for determining the location of the object in space through audition (see Appendix). Equation 4 shows the result of the derivation, namely the errors in sound localization that would occur with growth of the head as a function of where the sound is.

$$\theta 1 + \sin \theta 1 \; = \; \frac{r2}{r1} \; (\theta 2 + \sin \theta 2) \tag{4}$$

Specifically, it calculates the angle (in radians) at which the sound would be heard after growth ($\theta 1$) as a function of where the sound actually is ($\theta 2$), and as a function of the radius of the head before ($r1$) and after ($r2$) growth. When the head size increases by more than 1.5 times from a newborn infant to an adult (by a factor of 1.62), a nearby sound that is actually at a 45° angle would sound to be at about 73°

for an adult who had no adaptation and believed himself to have the head size of an infant. For small angles and close sounds, θ is approximately equal to sin θ, so this function is linear, like the rules of adaptation.

Since space is growing short, I will limit myself to mentioning only one thing about this equation. There would be no errors for a target straight ahead (0°), regardless of the amount of growth, and maximal error off to the side (90°). This is notable because for pointing, it was nearly the opposite; there is no error with growth at 90°, but a large error at 0° (Equation 1). This tradeoff between the two sense modalities might be useful for figuring out which modality is wrong when growth causes them to disagree.

There is no particular reason for thinking the mind "cares" whether a detected internal error is caused by movement, growth, or anything else—an issue also noted in a previous chapter. Despite a closer match of movement than growth to the linearly-biased adaptation mechanism we possess, adaptation should still help correct any detected internal error, regardless of cause.

Lost On Campus

The final topic of the chapter takes us back to the first topic of the chapter, a posttest to the pretest, if you will. Between Old Main and Social Sciences on the University of Arizona campus there is a nice expanse of open space. Hmmm, what could we do with all that space...Subjects were lead from a position we had them start from to a new position. Followed by a brief pause, they were then led to a third position. Subjects were now asked to walk back to the start position on their own and stop when they got there. We also asked them to turn to face the same direction they were in at the start. Since we did not want subjects to do the task just by seeing where the start box was, the catch was that they were blindfolded or had eyes closed through the whole task, from start box to finishing position[13]. Note that subjects have to generate a new route to find their way "home", so it requires creating some kind of spatial map. It's too bad we could not also eliminate cues like sounds and texture variations in the ground for a purer map-making test.

Figure 9-10 shows where each of 55 subjects arrived when trying to find the start position. Many people did acceptably well—remember, no one had the benefit of vision either to form the spatial map or to navigate back home. Three people were especially close with the best subject finding the *exact* location in the horizontal direction and only 6 inches away in the horizontal direction! Four additional people appear to cluster nearby, although the cutoff for being "very good" is arbitrary. Thus, there seem to be about seven super locaters, nearly 13% of the students, who had only small under or overshoots of the correct location in one or both directions. The majority of people made larger errors. For example, three people walked along exactly the right path, but stopped too soon. They can be seen along the dotted line of the third triangle leg that connects the the third point to home. A couple of others were on that path as well but then just kept right on going and overshot the right location by a considerable amount. The most common error was to head in a direction that did not have a sharp enough angle. Some of those were pretty close in distance, others not so much—perhaps somehow sensing that it wasn't quite right and just kept on walking. You can see many of those in the lower left quadrant of the plot. Not everyone walked in one straight line , even though we asked them too. Many of the errors seem to be sensible, especially given people could not use vision.

However, I think it is fair to say that at least 10 people were lost. Imagine you were doing the task and have just been lead to the 3rd point from the 2nd point. Home is to your right and behind you. Can you feel that? Yet some people headed off in a direction to the left instead of right (6 people—on the plot these are to the right of point 3), or forward instead of backward (4 people—look below point 3). One additional person headed in the right way but is so far off it's hard to describe that as anything but lost (the upper leftmost point in the plot). In fact, that person had to be stopped at the edge of the campus green, about to collide with obstacles. Of people who went left instead of right, one person interestingly seemed to get the angle correct and was even reasonable about distance, but with the direction reversed (they went both left *and* forward). Another seemed to be on a path that retraced steps back to the position they had just been lead on. Given some of these errors, it would have been interesting to have asked people how confident they were that they

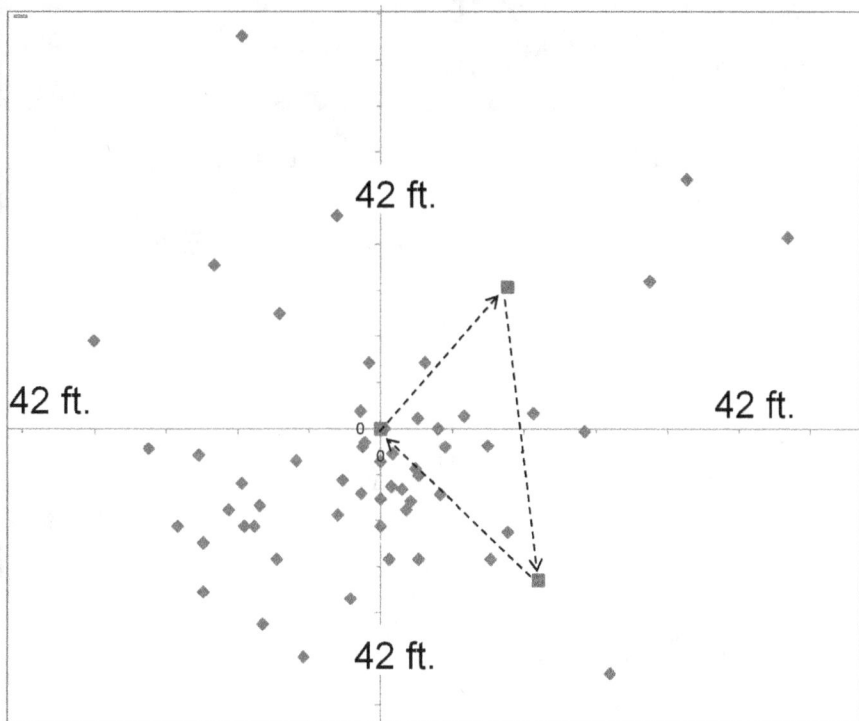

Figure 9-10 Results of the navigation study. Subjects were led blindfolded on a path outdoors (dotted triangle). They were asked to walk a straight line back to the start position (center; 0, 0) on their own while still blindfolded. Each small diamond reflects one subject's arrival location. (Look closely for the nearly perfect subject. The location is hard to see because the diamond falls within the marker for the start position!). Annotations are shown on next page where triangles refer to male subjects and circles to female subjects. Lost subjects indicated with labels.

Figure 9-10 continued Male Δ, Female O, and Lost subjects indicated.

had arrived back home. Did these people think they were right or did they know they were hoplessly lost? Did they feel disoriented? The furthest person in the wrong direction was 71 feet away and would have kept going if not stopped. Another person may have also strayed further, but was our second person that had to be stopped because of impending collision (with rocks). We were not expecting anyone to be that far off. One can only imagine how they would do with something more complicated than our triangle of three positions.

Here's a good question: Are the people who got lost the ones who predict they would get lost? We did something sneaky here. People who said they got lost (see survey results in Table 1 at the beginning of the chapter) were invited to participate in the study. *But we didn't tell them that's why they were invited.* It was at least two months after they filled out the questionnaire and they had so many questions to answer from so many different researchers. It is unlikely that they would make a connection between one answer from several months earlier and the invitation to this study, which we also did not announce as a "space" or "getting lost" study. I am emphasizing this because I think that results might well be different if people are made aware of their own self-assessments of good or bad navigation ability right before they are asked to do a task that requires navigation ability. By the way, although I was prepared to answer questions about why they were invited (something misleading but harmless), not a single person asked. It would appear that college students with five or six classes, part time jobs, midterms, and a requirement to be a subject in multiple experiments have other things on their minds.

The answer is that there did not appear to be any relation between those reports of getting lost and actually getting lost. Subjects that got lost had both people who reported at least one spatial difficulty and people who did not. They were in the same proportion as those who were super locators and those who performed averagely. Yet about the same percent of people get lost as self-reported (See Table 9-1)—just not the *same* people. Clearly, we have much rather interesting work to pursue.

Surely you are wondering about the gender of the subjects, especially the Losts and the Founds. Of the 10 who were lost, 8 were women and 2 were men. Wait! There were more women than men in the study—32 women and 23 men. Still, that is a higher percent of women than men getting lost—25% vs. 13%—nearly double. This is suggestive of a gender difference in actual performance, not just self-report, but a great deal more subjects would be needed to confirm this. Of the Founds, 2 out of 3 from the top set of super locaters were male, but the remaining 4 from the close "very good" set were female. There is also a great deal of overlap generally between the performance of men and women. Figure 10-11 shows all the subjects coded by gender.

It is hard to bridge the gap between getting lost in space—or *not* getting lost in space—and keeping spatial localization abilities accurate through the process of adaptation. It is easier to take things apart than to put them back together.

Adaptation is our second major type of perceptual learning in which experience causes a change to perception. We have looked at a number of properties of adaption, including the conditions necessary to produce adaptation (disagreement between systems, singe source assumption, a priori constraint on perception), the rules of adaptation (largely linear, rigid preferred, may follow transformation geometry), how learning manifests in behavior (e.g., change in pointing), and the purpose (error detection in body schema with movement, drift, but don't completely rule out growth). There are a number of take-home messages one could offer about this important type of perceptual learning known by the unfortunate too non-specific (and dull) name "adaptation". I will highlight just one. *A perceptual system is kept accurate by comparing it to another.*

I warned you that you would hear more about prisms than you ever wanted to know. After this bonus double-length chapter, you can take the prisms off now.

When is it? Can Time Adaptation Change the Perceptual Fabric of Space and Time?

"I am afraid I cannot convey the peculiar sensation of time travelling. They are excessively unpleasant. There is a feeling exactly like that one has upon a switchback—of a helpless headlong motion! I felt the same horrible anticipation, too, of an imminent smash." The excerpt is from H.G. Wells famous 1895 work *The Time Machine*, p. 16. The Time Traveler is describing a journey into the future.

"Inside this...body I grow old but outside—in every part of me but my mind and soul—I grow young." A quote from *Confessions of Max Tivoli* by Andrew Greer in 2005, p. 5, in which the main character passes along the time dimension backwards, from old man to young boy.

An old cup that is gathering dust on the table can be seen (or felt or heard) in one place but then can be seen (or felt or heard) in a different place following an appropriately chosen experience. Adaptation changes the perception of space, as seen in Chapter 9. Time is the other fundamental medium within which everything is perceived. Can your perception of time be changed as well? This seems hard. In space, we move things, including ourselves, around physically at will from one place to another as if space were sitting there waiting to be filled. Moving things around space the same way perceptually as we can do physically makes intuitive sense. Do we also move things around in time? Do we have that kind of control over time? One way to approach the question is to think about all ways in which the perception of time seems altered just in the course of everyday life.

Time Bandits

Time flies when you're having fun! When you are doing something that you really enjoy, it seems like you are robbed of time. Maybe an hour goes by but you would have guessed it had been only been fifteen minutes. The activity just started when it suddenly comes to an end. Another sense of the popular saying is being so completely absorbed in what you are doing that you lose complete track of time. *Flow* is a term that comes from Positive Psychology, a newer movement that emphasizes optimal mental health. In a state of flow, all attention is focused on one activity and everything else is lost— nagging thoughts, feelings of hunger, negative emotions, and a sense of time elapsing. Now consider the opposite of "time flies when you're having fun", although there is no catchy little phrase for it. If you are doing something that you would rather not be doing, perception of time is also altered. Waiting in a doctor's office feels like it takes even longer than it actually does. Lectures are boring—not to deliver them, just to listen to them. We like to hear ourselves talk and it's dreadfully boring to hear somebody else (tip: ask questions to stay awake!) As a result, time seems to slow down. Time seems to move more slowly. "Gawd, isn't it an hour and fifteen minutes already?" Plenty of clock-checking occurs. My former colleague, Paul Bloom, said he liked going to boring colloquia so that his life would feel like it lasted longer. There may even be an opposite of flow: an unusually heighted attention to the passage of time when trapped in a disliked activity. So, the same hour and fifteen minute block of time can either be perceived to be very fast or very slow, flying or dragging, depending upon the activity.

Now think of an emergency. Different experiences about the passage of time are reported during episodes of unexpected negative events that produce intense emotion. More time may seem to have elapsed than it really did: "I saw the tree branch falling and I don't know why I just didn't move out of the way". He didn't just move out of the way because it happened ridiculously quickly and only seemed like it happened slow enough to allow escape. Emergencies allow for many more thoughts than one would otherwise have in a fixed duration of time. Undoubtedly, the biological fight or flight instinct is the adaptive response that readies the body and mind for the swift action that an emergency requires. But regardless of the biochemistry involved, the perceptual consequence of the heighted arousal and all those crammed-in thoughts is an altered sense of the amount of time that has gone by. The description that one's entire life flashes in front of the eyes may be a similar time-altering experience during moments of extreme duress. Opposite reports also occur in which even a long-lasting emergency "happened so fast" that much of the time cannot be accounted for. Again, we have a perceptual acceleration or deceleration of time, though exactly which will occur during an emergency situation is uncertain.

Sensory deprivation tanks were once a well-known option for relieving stress. A person (voluntarily!) got into an enclosed, dark, soundproof tank where he or she floated in salt water the same temperature as his skin. An environment without anything to be seen, heard, or felt seems more like a perception experiment than a medical treatment. What happens to your thoughts when perceptual input from the world ceases? "I thought you forgot about me!" The session was actually only an hour long but could seem to last hours. If you are shut off from external stimulation, perception of time can change. Or consider what may be the opposite of too little stimulation: going somewhere new. You drive along a new route, finish whatever has to be done, and then drive home by the same route. Which ride seemed faster? Everyone says the ride coming back seems faster. Perception of time is different for the outbound and inbound journeys even though objectively they take the same amount of time, assuming that traffic delays can be equated. Was the outbound journey the less enjoyed activity? Was it crammed with more thoughts? We are not trying to analyze the events, just collecting as many instances as possible in which the perception of time seems to change.

Another notable example comes from drug intake, both legal and illegal. Cocaine causes intervals of time to be overestimated; that is, everything around you seems to be going so slo...w...l...y. Cocaine and other stimulants that affect dopamine levels in the dorsal striatum of the brain, such as Ritalin, cause increases in the amount of time activities feel like they are taking while depressants may do the opposite. Other psychoactive drugs, such as LSD, produce reports of unusual time distortions, like "time stopping", or a merging of past, present, and future, or disruption to the order in which events seem to occur.

Changes in time perception can also be comprised of just simple shifts earlier or later. Switching to Daylight Savings Time takes place in most regions of the United States and means having to shift your perception of time by an hour. At first, you may rely on the clock to know when it is time to eat or wake up but soon your internal clock is shifted. Once it shifts, what feels to your body like early evening or time to wake up will now be objectively at a different time of day. A more dramatic example of shifting time perception forward or backward occurs after flying to another region of the world that has a time difference of several hours. After a few days, perception of time shifts and you can now fall asleep easily when everyone else does. Incidentally, not all internal clock changes are artificially induced. The sun is going to rise at different times of the day during different times of the year. This requires a change in your body's perception of the start of the day for the same "objective" time on the clock. Change in the number of hours of sunlight per day throughout the year may require internal clock changes. You might have been born in the North Pole with all daylight half the year and all darkness the other half.

Different kinds of experiences leave us with no idea of how much time has elapsed. Sometimes following very deep sleep, people are clueless as to how much time has been spent asleep. You can actually trick the newly awakened by setting clocks to provide false information. The concept of flow described earlier may have the same loss-of-time effect. If you have ever been to a casino, you may have lost track of whether it was day or night even. Incidentally, this is not an accident. Casinos are intentionally kept continuously bright and in some of the bigger places, extra oxygen is pumped into the air to keep people alert as if it were daytime. The hope is that by confusing perception about whether time is passing, patrons may continue to gamble and spend more money. You might also lose track of time when you are sick or have a concussion. If you were in a coma, certainly you would lose track of time. Anesthesia can leave people with uncertainty over how much time has elapsed, though more frequently with a lost block of time. The hours seemed like seconds and when the anesthesia wears off, patients are convinced they have not yet been to surgery. Ever get momentarily disoriented about time when you exit from a dark movie theater? Or consider jet lag from travelling across time zones before adapting to the time shift. With jet leg, different internal body clocks that normally agree provide different information. For instance, body temperature may be set as if it is the middle of the night whereas you are getting hungry as if it is the middle of the day. Not having a coherent sense of the time leads to confusion of mind and body.

There are quite a few examples of experiences which may involve changes to time perception but require more scrutiny. During REM (rapid eye movement) sleep, events seem to occur in dreams that would take hours in waking life yet can only be occurring in minutes. Or consider the common experience of *déjà vu*. It is an eerie, almost otherworldly, feeling that an event happening now has already happened before. Sometimes it can seem so strong that you think you can predict what will happen next. That is an odd intermingling of the sense of past, present, and even future. Incidentally, this is a fourth example of a perception phenomenon that is occasionally attributed to a supernatural cause. Déjà vu has been said to result from remembering an event from a past life. The issue of déjà vu also leads to considering just the ordinary memory of past events. Could it be argued that recalling any event is a type of going back in time, a mental-time travel from present to past or an experience of simultaneous past and present?

We will consider just one more normal change to the perception of time. Imagine that you are interrupted from an activity and asked to estimate how long the activity lasted. How do you think you would do? People interrupted from an activity after 1 second, 5 seconds, 1 minute, or 20 minutes can reasonably estimate times but are nowhere close to perfect. Yet if you practiced and repeatedly estimated durations following interruption, you would become very precise. The perception of time not only can speed up, slow down, shift, get lost, or disordered—it can also get more precise. If this sounds familiar, it should. Chapters 5, 6, and 7 discussed the other perceptual learning process, perceptual expertise, which can improve the ability to discriminate similar stimuli of any sort. Whether there are similar wines, faces, X-rays, sheep, fabric, or temporal durations, perception will change and become more precise and less variable as expertise is acquired.

Finally, as there were with space, there are abnormal changes to time perception. Dementia patients may have no idea what time of day it is, may feel the past and the present co-exist at the same time, and may not be able estimate the duration of events. Schizophrenics have time perception difficulties that can involve uncertainty as to the order in which events occur (and perhaps, therefore, contributing to delusions). The neurotransmitter dopamine is involved in schizophrenia, as it is in Parkinson's disease, the latter of which also has reports of time perception anomalies. People with attention deficit disorder may be poor at estimating durations and have a high variability in their time judgments. This can lead to overestimations, especially for short time intervals that are perceived to be too long (and perhaps therefore much too long to sit still). In all cases, it would be interesting to assess the extent to which changes in dopamine levels go along with changes in time perception. As with space, loss of accurate time perception is disruptive to normal functioning.

On a small time scale, inaccurate perception of when something is occurring means not being able to successfully cross the street or catch a ball. On an intermediate time scale, it means not knowing if it is morning or evening. And on a long time scale, it is not distinguishing past, present, and future.

Depending upon the individual experiences you have had, each example of a time anomaly mentioned here may have seemed compelling to a greater or lesser degree. But with so many examples, the idea of changing the perception of time may start to feel a lot more intuitively plausible than it did at the outset. In fact, these many examples suggest that it is a lot easier to think of time perception getting altered in everyday life than it is for space perception. This, in turn, suggests that time adaptation ought to be at least as easy to obtain in the lab than space adaptation. However, most of the experiments in adaptation have been done on space, not time. Indeed, I know of only two laboratories that have attempted to get genuine time adaptation analogous to space adaptation. Neither succeeded in proving that a change in time perception occurred, although I do think both may have successfully produced such change.

Thinking About Time

If time is more difficult for everyone—including researchers—to reason about than space because it does not stay where we left it or because we cannot move freely about within it, then I suggest we use Einstein's trick of drawing an analogy to space. Space is an easier domain to visualize and reason about. Time can be translated into space, reasoned about there, and then mapped back to time after obtaining an answer. This is easier than it sounds and will avoid errors in reasoning about time adaptation that have occurred in the past. One hundred years of space adaptation research should make it even easier. This is not just an abstract exercise because, to preview, successfully changing the perception of time through adaptation would have an astonishing implication that is not found in space adaptation: Genuine time adaptation implies a perceived reversal of cause and effect! If you never quite grasp this you will be in good company because several of my colleagues remain incredulous that this could be the prediction. However, if you suspend what you think you know and follow the logic, it will be easy to see the inevitable conclusion.

Time Survey

We will turn shortly to the issues of drawing the analogy of time to space, followed by considering attempts to get time adaption, and finally, deriving the conclusion of cause-effect reversals. First though, here briefly are some results on the time questions from our survey of college students involving space and time introduced in the last chapter. They are not all directly related to the current discussion, but you may find the results interesting anyway. Self-reports of time anomalies are greater than space anomalies. Men and women seem more equally affected than for space, except perhaps on being late for appointments.

Table 10-1 Space, Time, and Well-being Survey

	% of subjects who answer "True"	
Time questions	Male (967 subjects)	Female (1,464 subjects)
Time always seems to me to pass more slowly or quickly than in reality	44.4%	45.6%
I am often unsure of when things happened in past	31.8%	32.5%
I have a tendency to be late for appointments	18.6%	23.2%
I am wearing a watch right now	20.2%	15.9%

Data collected from 2008-2012, Bedford, University of Arizona. Other questions in Table 9-1.

I have been remiss in issuing a reminder to marvel for this topic. It is actually pretty amazing how many experiences many of us have in which time seems to slow down or speed up or undergo other alterations. I find that people often are biased into thinking that time perception is stable, despite their own experiences to the contrary.

Analogy of Time to Space

Beginning with differences between space and time, space has three dimensions but time has only one. Left-right, up-down, and in-out characterize the three independent dimensions of space. Run left, now run to the right, jump up, fall to the ground, now run in front of you a few meters then back to the start and you will have moved within all three dimensions of space. There is just one dimension of time within which to move, characterized by elapsing seconds, minutes and hours rather than centimeters, meters, and kilometers. The second notable difference is that spatial dimensions are symmetrical but the time dimension is asymmetrical. Despite the effects of gravity, things can be placed up above as well as down below, you can walk to the left as well as to the right and if you move forward in space and then if you want, you can turn around and retrace your steps and walk back to the starting point. If only this were true for time! Except in our creative works of fiction and our deepest desires, we cannot go backward in time. The one dimension of time has two directions, backward and forward, past and future, but unlike space we cannot move in both of these directions. We can only move in one direction along the time dimension.

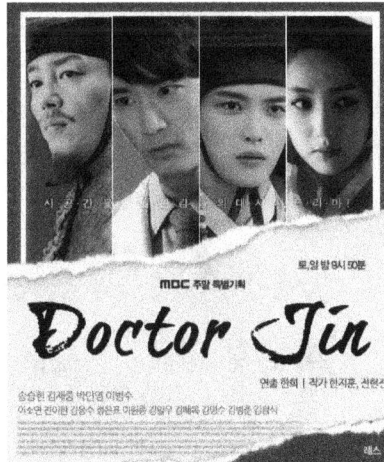

Figure 10-1 Going back in time is a popular theme in entertainment. In Korean television drama *Time Slip Dr. Jin* (right), a doctor is transported back 150 years after removing a teratoma from a patient's brain. In real life, social psychologist Ellen Langer's experiment mentally brought senior citizens back 20 years with an environment containing only items from back then.

Despite these rather noticeable differences between time and space, there is still a useful analogy to be drawn between them. Note that they are both continuous dimensions. Pick a spot here and another one, say 12 kilometers from here, and every single spot in between these two also has a spatial location, such as 1, 7.45, and 10.21345 km. The same is true for time. Pick any 2 time points, like now and 12 hours from now, and every single moment between them has a value in time. Time and space are also both ordered dimensions in which any value can be identified with its place in the sequence. For instance, 5 kilometers is further along on the space dimension than 4 kilometers which is further than 3 kilometers as 5 hours later is further along on the time dimension than 4 hours which is further than 3 hours.

For the analogy then, even though there are three dimensions of space and only one of time, pick just one continuous ordered dimension of space to map onto the continuous ordered dimension of time. I will pick the left-right dimension of space, despite a conceivable argument that gravity may make the up-down direction more analogous because it asymmetrical like time. The reason for the selection is that most of the space adaptation research has been conducted on the "horizontal" left-right azimuth direction of space and because it is a spatial direction that we are especially good at perceptually. The analogy:

Left in space = Earlier in time

Right in space = Later in time

Straight ahead in space = Right now in time

If something is earlier in time then that corresponds to something being further left in space. If something is occurring later in time, that corresponds to it being to the right in space The earlier in time something occurs, the further to the left it is in space and the later it is in time, the more rightward. Considering the space somewhere in the middle of left and right we have a position that is straight ahead such that everything else is either to your left or to your right. What corresponds to the midpoint in time? It would need to be something such that everything on one side is earlier and everything on the other side is later. The midpoint then is right NOW, something that is happening right now, right this second. Well, we just lost it again because now that time is the past already, corresponding to something leftward in space. For the analogy, something which is happening right this second is the equivalent in space of it being straight ahead. This serves as the midpoint of the time dimension.

Besides using a mapping between time and space to allow reasoning about difficult problems within space instead of time, space and time have a natural connection. They are fundamentally interconnected because it takes time to traverse space. I am sure you have encountered mention of the combined four dimensions of space-time or the space-time fabric, manifold, or continuum. In addition, time and space are quantities both in the world and in perception from which there is no escape; everything must be done within these two mediums of existence.

As for the big asymmetry in the time dimension but not in space, it is that difference that leads to the peculiar predicted consequence for time adaption of perceived cause-effect reversals. It is also where converting from time to space will be maximally useful.

Time Adaptation

If there is a discrepancy between sense modalities, recall that adaptation is a reflection of a long-term resolution to the disagreement such that one or both of those modalities are changed in order to fix the problem. To get space adaptation, the two modalities disagree about the location of an object. One modality determines that it is in one place, another modality indicates it is somewhere else, and one or both modalities are changed in order to restore the accuracy of perception. For time adaptation to be successful there must be a discrepancy between modalities, not about the location of an object in space but about when something is happening in time. Arguing by analogy to space, if we can create a discrepancy between two different modalities about when something is occurring then one necessary step for adaptation will have been met. Hopefully, we should get a long term resolution to the discrepancy which would involve changing the perception of time within one modality or the other. That's basically what the time experiments try to do. Before analyzing time adaptation more closely, as well as whether the laboratory attempts were successful, consider what was actually done in the laboratory experiments.

Experiment

Douglas Cunningham has one of the labs which have attempted to get time adaptation. What he and colleagues did for one of their paradigms[1] is to have people controlling an airplane—not a real airplane, which would be even better, but a pretend airplane on a computer monitor. Subjects moved a mouse on a mouse pad and a picture of a plane would move on the computer screen. Subjects only had control of the left and right directions. If they moved the mouse to the left, the airplane would move to the left, move to the right and the airplane would move to the right. They were not given control of the vertical direction; instead, the plane would start at top of screen and would drift downwards. As it drifted downward, it entered a field of obstacles. If the plane was not moved left or right, eventually it would crash into one of obstacles (see Figure 10-2). The subject's task was to keep it from crashing into the obstacles by moving it left and right at the appropriate times.

The experiment consisted of three parts, pretest, training, and posttest. Pretest and posttest are when the subject is tested while training is the experience provided to try to obtain time adaptation. The effects of training are assessed by comparing pretest to posttest; if there is a difference between them, it is concluded that the change was a result of the training. The experimental design of pretest,

Figure 10-2 Subjects in the time adaptation experiment saw these obstacles on the screen. They had to maneuver a virtual airplane left and right to avoid collisions. Cunningham, D.W., Billock V.A. & Tsou, B. H. (2001) Sensorimotor Adaptation to Violations of Temporal Contiguity. *Psychological Science 12(6,)* 532-535. Published by Association for Psychological Science (APS); Reprinted with permission.

training (also known as exposure), and posttest comes, of course, from space adaptation studies. In the last chapter, we saw this classic design as well. Cunningham and colleagues explicitly tried to model the time adaptation experiment after the many space adaptation experiments and they did get this part right.

Turning to what they did for each of the three parts, testing before training (pretest) involved having the plane start descending automatically and continuing to do so at a fixed speed. They used a bunch of different speeds in random order—10 different ones—and they gave each subject each speed 5 times for a total of 50 trials. They wanted to see how well people did at avoiding the obstacles at the different speeds. If the subjects crashed into an obstacle, it ended the trial. If they made it all the way through to the bottom, that also ended the trial. After all 50 of these trials were complete, the experimenters noted the highest speed at which each subject was able to successfully get through the obstacle course on at least 4 out of the 5 repetitions. The higher the speed, the harder the task, so this highest speed pretest served to determine how successful subjects were at performing the task before any training.

For training, they introduce a time delay. This is the fun part. By time delay, I mean there is a lag between when the subject moves his or her hand and when the airplane is seen to move on the screen. Move the hand to the left and airplane moves to the left—but not immediately. It takes a while before it shows up on the screen. Think of how frustrating that would be. If you ever had a slow computer internet connection, you may have experienced the frustration that occurs when there is a delay in seeing the results of your actions. Or consider your favorite video game and imagine that there is a delay. You move your body to the left but then don't' see the avatar on screen move for a while. Think about what that might do to your performance. Indeed, subjects complained at the beginning of training that they would not be able to do the task with the delay and were told to just do the best they could. The time lag introduced was 200 milliseconds (msec) which is only one fifth of one second (1 second = 1000 milliseconds). A small fraction of a second seems like it would not be of any consequence but when engaged in visual-motor control, the delay is significant. We see results of our actions much faster than this. If the delay were 20 msec, it would not be noticeable, but increase the delay to 200 msec, and it is rather noticeable and disruptive.

Although the experimenters always used the same 1/5th of a second time lag during training, they did try to get people to adjust to it gradually. The speed at which the airplane descended was started at the slowest setting. Incidentally, note how this is another example of easy-to-hard training discussed for a different perceptual learning process in Chapter 7. In this experiment, they took the 10 different speeds used during pretest but instead of randomizing them, the slowest speed was presented first. If a subject succeeded on making it to the bottom of the obstacle course on 8 out of 10 trials at the slowest speed then they would step up the airplane decent to the next fastest speed. If a subject collided with an obstacle, they repeated the trial at the same speed until he or she succeeded at 8 out of 10 trials. Mercifully, if a subject hit an obstacle 10 trials in a row then they would just stop the training. Or if a subject just could not succeed 8 times, they finally stopped the training after giving 70 tries at that speed. Subjects who met the criterion were given the next highest speed and the procedure repeated with each higher speed introduced each time the subject cleared 8 out of 10 trials. This may give some sense of how much training subjects got with the time delay which would be different for different subjects. Note that subjects can have quite a few trials of training with time delay trying to meet the criterion at each speed and subjects who do very well get to practice with the time delay at even higher speeds.

When training is over, subjects are tested again (posttest). Posttest should be identical to pretest to ensure that any changes that result are due to the training itself rather than to a different test procedure. Theirs is a little different, but close enough. The different airplane speeds were presented again as in pretest and the highest speed again determined at which subjects successfully got thorough the obstacles on 4 out of 5 trials. Since posttest needs to be the same as pretest and there was no lag during the pretest, the lag was removed for posttest. Following training, the mouse was returned to its normal state and now the mouse/hand movements were effectively immediately presented on the screen. Subjects tried again to avoid colliding with the obstacles by moving the mouse left and right without a time lag, as they did in pretest. The highest speed attained served as the measure for how well subjects performed the task after the training.

The results of the experiment included that subjects stopped complaining about the time lag. When they persisted with trying to avoid obstacles despite the time lag they became able to do the task. Many people did quite well during the training phase of the experiment despite thinking they would never be able to. But, of course, we want to look at how their posttest data compares to pretest. What Cunningham and colleagues found was that subjects did much worse than they did before training. When the time lag was returned to normal, the highest speed at which they could get through the obstacle course was lower than the speed attained before training. Having just gotten practice with the time lag, they now did poorly without the time lag. If you get used to a time lag then it is the absence of a time lag which becomes difficult for you! Subjects also did worse than subjects who were given an unrelated task to engage in during the training session without a time lag.

Analysis

The authors concluded that the experiment showed genuine time adaption. Their argument—and note, I did not say I agreed—was two-fold. They noted that people did get better at dealing with the time lag and even more importantly, that they had an aftereffect from the time-lag training. The aftereffect was that subjects' motor performance was worse when the lag was removed. They argued that this is analogous to space adaptation, in which an aftereffect following prism exposure once the prism distortion is removed demonstrates that adaptation has taken place. Whenever there is an aftereffect following training, they argued, this shows that people had successfully adapted to time changes/lags.

Did they succeed in showing genuine time adaptation? To jump to the end, their data did not show that. Here's why. Using our analogy between space and time (right = later, left = earlier) is helpful here. Recall for genuine space adaptation, there has to be a change in where the object is perceived. A prism that shifts the image to the right by 10° causes an object to feel to be in one place but look to be in another place further to the right. If adaptation brought on by the disagreement is successful then either the perceived visual position shifts to the left or the perceived proprioceptive position shifts to the right. By analogy, a time lag that delays the image by a fifth of a second causes an event to be perceived through the hand at one time but look like it's occurring at a different time a little later. If adaptation brought on by the disagreement is successful then either the perceived visual time shifts to being earlier or the perceived felt time shifts to be later. Therefore, in order to succeed at showing time adaptation they would have to have shown that perception of time had actually changed. There needs to be a change in either when vision detects the event to be occurring (e.g., the airplane is seen to move sooner than it does on the screen) or when proprioception detects the event to be occurring (e.g., the hand is felt to be moving the airplane later than it actually did). The experiment did not show that either of these changes, or any change to time perception, occurred.

What the experiment showed instead is that people got better at performing a task when they had practice. But that's not the same as adaptation. Getting better on a task is not same as changing perception. People get better for all sorts of reasons. Think of any motor skill that you can do. You got better with practice. This does not mean that you now perceive the world differently. Motor skill learning need not involve changes in actual perception and is, in fact, is a different kind of learning than perceptual learning (See Chapter 12). People get better at nearly everything and not everything is perceptual learning. Getting better is clearly not sufficient for showing a change in perception.

But what about the aftereffect they found? Cunningham and colleagues also showed that people then performed worse when they took away the time lag that they had practiced with. When the time lag was removed, subjects were worse than if they never had any training at all. This consequence does not show that there was a change in perception either. Suppose I gave you a long sequence of letters to type repeatedly and it was always the same:

wolkknerpAmmvbytrqoiHGubpl

At first, typing a string of random letters accurately may take longer than typing familiar words. But if you were motivated enough to do it over and over again—say I give you $500 if you cut your time in half —you would become proficient at typing the sequence. Now suppose I changed the sequence to one that uses the same letters but in a different order:

owklmybermkvtprAoiqGHublpn

How do you think you would do? Not only would you type the

new string poorly, but you would do worse than if you had never gotten training on the initial string. Getting good at one task now interferes with a different, but similar, task. That's not perceptual learning either. Now, had practice with the initial string caused you to see *Working Am by Tranquil Guppies* instead of *wolkknerpAmmvbytrqoiHGubpl* or made you feel like you were typing an "r" when you were really typing an "l", that would be a change to actual perception. Otherwise, just improving at a task and then performing worse at another task does not imply a change to perception. The time experiment showed that people got worse at a task after practicing with a different task but did not show adaptation.

An anecdotal example occurs every year at the University of Arizona when its basketball team, the *Arizona Wildcats,* plays a charity event *Lame for a Game* with a team of experienced wheelchair-bound players. The height of the net in wheelchair basketball is proportionately closer to the ground. Despite the great expertise of the Wildcats at basketball, they do not do all that well. These athletes appear to do worse than one would think even a bunch of casual basketball players might do if they tried basketball from a wheel chair for an afternoon. Proficiency at basketball includes getting exquisitely good at the exact height of the net. Change that, and we expect the experts to do worse at a different, yet similar task—in this case basketball with a different net height—than if they had never trained so extensively at that one height.

For adaptation to space in the last chapter, we saw that there were imposters. These were improvements to a task with decrements/aftereffects afterwards when the prism was removed that looked like they were adaptation resulting from the conflict, but were, in fact, something else. For instance, intentional "world learning" looked like adaptation but was not because the perception of location remained unchanged. The Cunningham and colleague time experiment assumed that the improvement and subsequent aftereffect of training were adaptation when it could have been other things, such as motor skill learning or conscious adjustment.

It is nonetheless intriguing what some subjects said during the experiment. They volunteered that as training with the time delay progressed, it now looked like plane moved at the same time as their hand moved, not later. Moreover, when the delay was taken away, "the plane seemed to move before the mouse did". These are the best pieces of evidence in the experiment that perhaps the perception of time did change after all. That is why at the beginning I said that the experiments unofficially may have succeeded in producing time adaptation, at least in some subjects. But these are anecdotes and we already have plenty of anecdotes that the perception of time can change, as seen in the first half of this chapter. The experiment unfortunately did not go beyond anecdotal subject reports to provide an objective demonstration that the perception of time had been shifted.

Cause-Effect Reversal?

I want to turn now to the most exciting issue of time adaptation: perceived cause-effect reversal. If perceptual adaptation is successful such that there is a change in actual perception of time—the hand feels like it is moving later than when it really is or the hand looks like it is moving sooner—then there is a striking consequence. When the time delay that is used to get adaptation is taken away, it should look like the plane (for example) moves before you ever move it with your hand! To make sure it is clear why, inspect Figure 10-3 which shows the time adaptation paradigm. Also compare this illustration it to the space adaptation paradigm shown in the last chapter in Figure 9-4. By following the analogy closely, you will see that perceived cause-effect reversal is an inevitable consequence.

Suppose you are a subject in the time delay experiment. During pretest when there is no delay (first row), your vision and proprioception/motor system perceive the event of moving the plane to be occurring at the same time. During training or exposure when vision is first delayed (second row, left panel), you accurately perceive the visual image of the plane to occur after you feel your hand move the

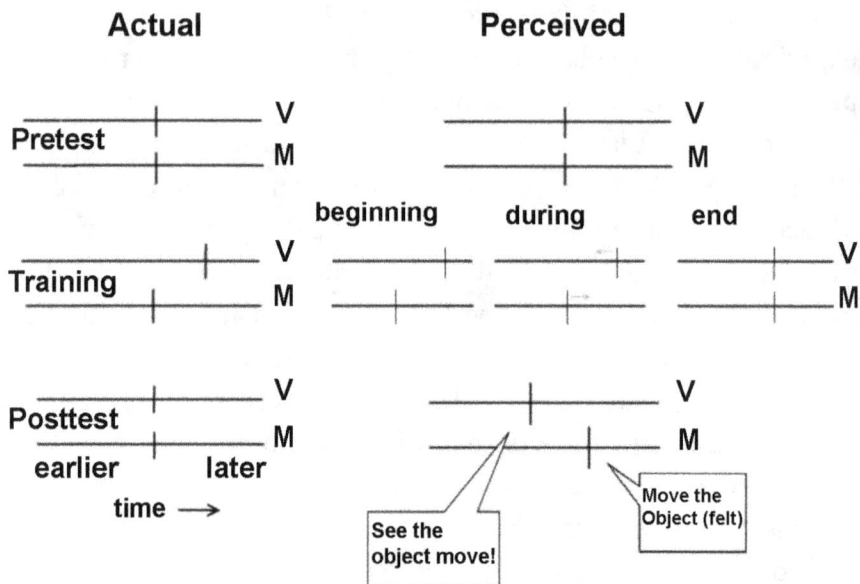

Figure 10-3 An illustration of the logic of time adaptation before training (Pretest), during training or exposure to a time delay (Training) and after training (Posttest). Vertical tick marks indicate when events occur on a dimension of time proceeding from earlier (left) to later (right). The left column shows when the events actually occur and the right columns shows when they are perceived to occur by vision (V) and by the motor system (M). Compare to the illustration of space adaptation (Figure 9-4) and note how the logic is identical. For time adaptation, successful adaptation implies that a perceiver will see the results of his action *before* he perceives himself to be causing the action (lower right panel). See text for details.

plane. Note how the tick mark for vision in the figure is further to the right (later) than the tick mark for the motor system. As training proceeds (second row, middle panel), genuine adaption will cause either vision to move towards the motor system's opinion, or the motor system to vision, or both, in order to fix the detected error caused by the disagreement. Vision will therefore either move to the left (earlier) or the motor system to the right (later), or both will occur. The arrows depict that the modalities have started to make these shifts during adaptation. At the conclusion of adaptation (second row, right panel) the shift is complete and one or both modalities have shifted. Shown are both modalities shifting and meeting in between. During posttest, the delay is removed (third row). As can be seen in the figure, the perceived visual image has been shifted to the left because of the adaptation that just took place and the perceived motor event has been shifted to the right. When the delay had still been present, these shifts realigned the modalities to the same place, as seen in the adaptation panel in the second row, but when the delay is absent, the shifts now cause a mismatch between perceived vision and proprioception. But what a mismatch!

Before proceeding, compare it to that space adaptation diagram (Figure 9-4) and note how the logic is identical. All the tick marks are in the exact same places. Whether in space or time, the imposed distortions produce initial disagreement between modalities, the adaptation phase then shifts one or both modalities towards the other one, and finally, when the distortion is removed, the adaptive shifts that occurred cause a mismatch between the modalities once again. As a subject, you are now adapted to a different environment, so reverting back to the original environment becomes a perceptual problem which would requires adaptation all over again. It is exactly the same for time as for space. Once you have verified that they are logically the same, look again at the last row for time adaptation, the posttest. Recall in the analogy and as shown in Figure 10-3, to the left in space is earlier in time and to the right in space is later in time. As shown in the last row, the event perceived visually is occurring to the left of the event perceived through the motor system. That is, vision's perception of the event is to the left, or earlier, than the motor system's take on the same event—yet the motor system is the cause of the event. Are you beginning to see the astonishing consequence?

It is probably already clear, but to unpack it in more detail (at the risk of making it sound more complicated than it really is) the event is moving the airplane with your hand. The motor/proprioceptive modality is the felt time at which you move the airplane with your hand. The visual modality is the seen time that you move the airplane with your hand....The visual modality of seeing the airplane move is now to the left of the proprioceptive modality of feeling the airplane move...That is, the visual modality of seeing the airplane move is now earlier than feeling the airplane move...The visual modality is also localizing the event further to the left, or earlier, than the event actually happens. So are you seeing the movement of the airplane that you yourself moved before you actually move the airplane?!

By analogy to space, the logic as applied to time clearly predicts that the effect will be perceived before the cause if time adaptation actually occurs. By perceiving the visual consequence before the perceiving the motor cause you would perceive the effect before the cause. It seems quite obvious actually when using the analogy to space. If you lose the intuition, just look at the time figure, especially the third row showing the posttest. Note that this does not happen in space because it does not matter which is to the left in space, vision or proprioception, since we can move freely in both directions. But it matters in time which modality is to the left of the other because we cannot move left—earlier—in time. Does this mean time adaptation is impossible after all because an effect cannot come before its cause? Not at all. Although you cannot see something before there is any-thing there to be seen, your perceptual system can think so. It just requires that the testing adheres to known principles of the physical universe.

You cannot test before an event actually happens and assess whether the subject sees anything. For instance, you cannot assess if the subject sees the airplane move as a result of his actions before he actually moves the airplane. But you can test afterwards. Once the subject moves the airplane, you can test when he feels to have moved the airplane. The felt motor position should be later than it actually occurred. For time adaptation, either the perceived visual event occurs sooner or the perceived motor event occurs later, or both. In Figure 10-3, both are shown as shifting. If we get lucky, time adapta-tion may have caused the felt motor system to shift, which is testable without violating any known principle of physics. But even a shift of perceived vision to earlier is testable. We can test the *memory* of it afterwards. The inevitable passage of time can now be used to our advantage. An instant of action is over in an instant. It becomes a memory very quickly. Memory and conscious perception often blur together. As long as a test procedure occurs after the actual events occur then perceptual changes can be detected that do not violate the laws of nature.

It may be worth trying to think of something in everyday life where a perceived cause-effect reversal occurs. One event that I can think of involved expecting a package to arrive later than it actually did. I ordered an item from the internet (shoes from Zappos) that was advertised to arrive in 3 days. The very next morning, the item arrived. When it happened, I had a peculiar experience in which it *almost* felt like the item arrived before I had even ordered it. Something about my time perception got a bit rearranged when expectations were violated which lead to an almost cause-effect reversal. In addition, when very small time intervals are involved, illusions frequently occur in which the events are perceived out of order—but probably you are not even aware that you have made a mistake. Finally, as noted earlier, schizophrenics and users of certain psychotropic drugs may get the order of events incorrect which would lead to perceiving an effect to come before the cause and not the other way around.

Thus, when we start looking, we may find examples of perceived cause-effect reversals, the predicted result of time adaptation. However, thus far, despite anecdotes and unofficial reports, neither time adaptation nor resulting perceived cause-effect reversals have been officially demonstrated in the lab.

Playing With Time Adaptation

While we wait for a clever methodology to prove time adaptation and perceived cause-effect reversals, discussion seems incomplete without thinking at least briefly about how one hundred years of space adaption knowledge might be applicable to time. Since we don't yet have a clear demonstration of the basic time adaptation effect, discussion of potential variants on the effect is admittedly abstract, so those who have had enough of this topic can safely skip it and put it off to some "future time". This is the kind of playing though that I greatly enjoy.

Conditions necessary to get adaptation

We know that a discrepancy between modalities is not sufficient to get adaptation. In Chapters 8 and 9, I raised the issue that there can be three reactions to a discrepancy, "oh wow", "uh oh", and "ho-hum". The third alternative refers to the conclusion that vision and proprioception disagreed about the location of an object only because they were referring to two different objects. If that occurs, there is no error of any sort and no action is required or taken despite the discrepancy. We also saw that there is a lot of perceptual deliberation concerned with the issue of deciding (not consciously of course) whether the two votes by perceptual systems refer to one or two sources in the outside world—the object identity decision. For time perception, too, adaptation should only be possible if the disagreeing modalities are judged first to come from the same source[2].

A time delay between performing your actions and getting to see them introduces the disagreement between the two modalities, the felt action and the seen action. You feel your hand moving an object, such as a plane, at one time, but see the object moving at a later time. In addition to this discrepancy, you would also need to conclude that you were seeing and feeling *the same plane* in order to get adaptation. Otherwise, you would be left with the "ho-hum" interpretation of the discrepancy. That would instead imply that you feel and see it at different times only because they are two different events. Two different events can occur at two different times just as two different objects could occur in two different places. There would no error, no need to do anything about the cross-modal disagreement, and therefore no observed adaptation.

Whenever displays are virtual rather than real, the perceptual ho-hum interpretation is a very real possibility. It is for this reason that I said at the beginning of the time experiment discussion that it would be even better if subjects were controlling a real airplane rather than a virtual one. For most computer-based virtual tasks, the hand controlling the action is one place, such as on a table, but the visual consequences of that action are somewhere else entirely, usually on the screen in a different location. In addition, your hand grasps and feels one object, such as the mouse or other controller, but you see a different one entirely—an airplane or a tennis racquet. Miraculously, you frequently still conclude it is one source despite all the differences between the two modalities. We know this because successful video game playing indeed involves overcoming these differences, concluding it's only one source responsible for all the input, and therefore feeling yourself erroneously to be where the screen is located. But if now there is a temporal separation imposed by the time delay as well as the spatial and shape one imposed by the basic video game setup, all these many differences between the modalities are increasing the chances that the impression of a single source will "break down". In this case, the object identity decision of one object or two will instead be that there are two objects/events, not one, and this assumed by the mind of the perceiver to be the explanation for the disagreement between modalities. That is, you mind may understandably conclude that you see the plane at a different time because your hand moving a mouse shaped object in one place at one time is, in fact, not really the cause of your seeing a plane shaped object in a different place at a different time. This would be the ho-hum conclusion for time disagreements and it could not lead to any adaptation.

This suggests that to successfully adapt to time delays, it is better to avoid displays in which one modality is already two steps removed from the other modality before even introducing the time delay. The other lab that has investigated time adaption did precisely this. Richard Held, the perception psychologist who has been conducting research for 50 years, and done nearly everything in perceptual adaptation, and therefore someone I have mentioned a couple of times already, now investigated time adaptation decades before virtual worlds existed[3]. Held had each subject look at his hand move side to side as he pointed to targets but introduced a delay in when the subject saw his hand. This task is not nearly as flashy (or fun!) as controlling an airplane but avoids problems with the object identity decision that could prevent adaptation. His subjects reported they felt their hands to be moving through a "viscous medium" at first, which then dissipated as training progressed. He found that if the time lag was made too great, the assumption of a single hand broke down and subjects no longer perceived that they were seeing and feeling the same hand. That is, capture of one modality over the other's opinion no longer occurred. Like the modern airplane experiment, Held did not include a direct test of whether time-delay exposure caused subjects to see their hand moving earlier or felt it moving later and thus we cannot conclude he showed genuine time adaptation either.

Content of adaptation

If the same rules of space adaptation apply to time then we expect there to be generalization along the time dimension the way there was generalization along the space dimension. That is, if a time lag procedure shifts one moment in time, we expect adaptation to shift to other moments in time, analogous to how training one spot in space generalizes to other locations along the spatial dimension. This seems sensible and almost like it would have to hold true for time. However, the analogy requires further scrutiny. Since time is always moving, is it possible to train with repeated trials of just a single spot on the time dimension the way we did for the spatial dimension?

Somewhat easier to apply to time adaptation would be the linear bias found for space adaptation. Recall that when my lab investigated a variety of different ways that space could be transformed, those considered to be linear transformations were preferred. Shifting vision with respect to proprioception by an equal amount at each position, or uniformly stretching or shrinking one modality with respect to the other, were the easiest transformations to adapt to.

They were also imposed by the brain under training conditions that were either ambiguous or more complex. These results predict that the easiest modifications to create for the perceived time dimension would be uniform shifts, expansions, and contractions, as it was for space. This idea holds promise because many of the common personal experiences we reviewed involve slowing down or speeding up perceived time, most likely uniformly. Uniform shifts like Daylight Savings Time are also easily accommodated.

In space, crazy transformations that violated topology were poorly accommodated. It is exciting to think about the equivalent non-topological distortions in the time domain. In space, one non-topological transformation explored was a many-to-one mapping in which three different visual locations in space were mapped onto only a single point in proprioception space. The consequences of such a mapping were that regardless of whether you saw an object at 10° to your left, 10° to the right, or straight ahead, you could touch the object by just keeping your hand in the same place for all three visual positions. For time, the equivalent is that three different visual times all correspond to the exact same felt time. Thus, if you caused an event with your hands at one moment in time, you would see the event occurring at three different times. It would repeat itself in time visually even after it was over. Imagine trying to adapt to that distortion!

How adaptation is manifested

Which modality should change if vision and proprioception do not agree about when an event occurred? In space, precision of the modality was relevant to which modality changed, as was the modality the perceiver was attending to or used as the lead/guiding modality. It would be interesting to apply these concepts to time. We know that vision is not that great at time, but we do not know if the proprioceptive concept of time is better or worse. Held's time-delay results in which subjects felt their hand to be moving in a viscous medium may be taken to imply that vision wins the conflict. This may suggest that vision is more precise than proprioception in time as well as space. This in turn leads to the expectation that vision would win over proprioception for a long-term resolution as well, not just for immediate capture where one modality dominates in the moment. Proprioception/motor system then would be the modality that changes in time adaptation, producing a change in the felt sense of when the hand is moved. By the way, if true, this is good news for testing procedures. Perceiving a felt action to occur later than it actually did should be a lot easier of a change to document than trying to test something that is perceived before it even occurs!

For time conflicts between vision and audition rather than vision and proprioception, we expect the disagreements would be resolved with changes to the visual modality instead. The dancing Christmas lights discussed in Chapter 8 reflected an auditory victory to a short-term conflict between vision and audition. Audition is the more precise modality at the apprehension of time. Suppose we were to design an experiment in which subjects conversed for an hour with a person whose voice was delayed by a fifth of a second. The visual perception of when the person was speaking should eventually shift to be later than it currently is. If the delay is then removed, the subject's perceptual system should detect that it hears the person before it sees him!

But keep in mind that precision is not the only determiner of which modality changes. It would be interesting to figure out how to manipulate attention or "guidance", the other big factor that influences which modality wins in space conflicts, for time conflicts and follow where the prediction leads. We also cannot be certain that space adaptation rules apply to time adaptation or that time adaptation doesn't have some unique tricks up its time equivalent of a sleeve.

Ending Time

There is a vastly open space—or should I say time—for time adaptation studies that have not yet been attempted. It seems as stationary as space, just waiting for someone with clever ideas, patience, and rigor.

It seems only fitting to end a discussion of anomalies in time perception with an anomaly. If past, present, and future can intermingle perceptually, and perceived time can even stop, then we do not have to end this chapter.

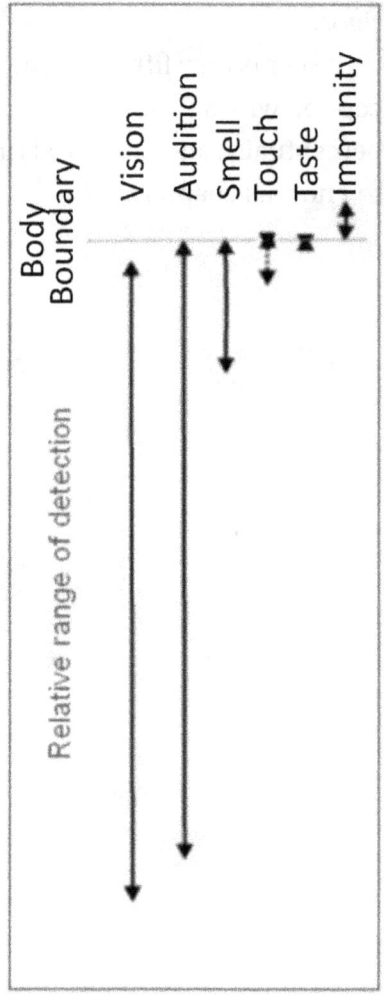

Figure 11-2 The immune system completes the sense modalities which together enable us to detect anything ranging from very far away way to very close to our core. It is such a perfect completion to the range of detection yet has been missing from the list of sense modalities. The immune system should be considered a sense modality like vision, audition and the rest. Figure 1 from Bedford, F. L. (2011). The missing sense modality: The immune system. *Perception, 40,* 1265-1267.

Meditation and Guided Imagery to Heal Illness: The Most Unusual Application of Perceptual Adaptation

Perceptual adaptation is a powerful way to keep your perception accurate. But the most powerful application of the adaptation mechanism was one that was not foreseen by the scientists who started adaptation research more than one hundred years ago. Ironically, the application was already present then, just not connected to adaptation—or indeed to science at all.

I am talking about using the mind to heal the body. A man's inoperable tumor shrinks after "laying on of the hands". A child's stubborn warts vanish after applying a "magic cream" that is just cream. A woman's intractable headaches stop after a pill that has no active ingredients. Breast cancer patients were claimed to have survived twice as long if assigned to a support group plus conventional therapy compared to conventional therapy alone. High blood pressure is reduced with meditation. Numerous cases of remission have been recorded through the years following visits to Lourdes or other healing springs. Clergy have "casted out demons" for as long as organized religion has existed. Tibetan monks can raise their body temperature practicing Tum-mo and entering deep meditation.

Mind over matter has been considered false medicine—quackery —and even false religion. At best, it has been ascribed to just a placebo effect. Would you mind if your cancer was cured by "just a placebo effect?"

Step 1. Mind-Body Interaction is Science

Using the mind to heal the body has been considered magical thinking, not scientific thinking. First, I would like to convince you that such influences are real science. That part will be easy because others have done the work for me. *Psychonueroimmunology* is the nine-syllable name given to the interdisciplinary field in science that emerged to study the relation between the mind, the nervous system, and the immune system—essentially, the mind, the brain, and the body. It seemed suddenly these things were being studied together, after so much time where connections between them were denied. What changed? It used to be thought the brain and the body each spoke its own language. On the one hand, chemicals such as dopamine were known to be taken up in the brain—neurotransmitters—to regulate mind/brain properties such as arousal. On the other hand, the immune system had a different system of modulation. For instance, interferon causes an increase in immune activation. Each of the two systems had a unique job to do and had its own specialized machinery with which to carry out its tasks. However, it was discovered that receptors in the brain were also responding to interferon and other seemingly specialized immune system messengers while at the same time, the immune system was responding to neurotransmitters. Two systems once believed isolated could communicate freely with each other. This meant that there was a plausible way in which a system critical for maintaining health (immune system) could indeed be affected by mental state (brain/mind). Mind-body interactions were starting to sound a lot less like magic.

Outside of the lab, ordinary people were turning to *alternative medicine* with increasing frequency. The physician and best-selling author Andrew Weil seemed like the opposite of a quack. He was trained at Harvard medical school, yet argued for herbs and meditation alongside CAT scans and angioplasty. His common sense approach of combining the best of all medical traditions appealed to many, even other doctors. They now go for training at his center for integrative medicine at the University of Arizona. For our purposes, I think he accelerated the acceptance in the public eye of psychological methods as a legitimate method for healing.

Step 2. Mind-Body Interaction is Perception

The next step is to argue that perception is important in mind-body healing. Up to now, the contribution of the field of perception to the emerging psychoneuroimmunology discipline has been non-existent. But remember, other things that have appeared at first to be supernatural have turned out to have a cause based in perception. There are two common psychological manipulations for bringing about positive physical changes that are frequently used together in mind-body medicine: *guided imagery* and *mindfulness meditation*. For both, perception is key.

In guided imagery (sometimes *visualization*) the mind is directed to intentionally invoke images. Imagination is used to conjure places, objects, or events that are not externally present. The goal is to influence psychological and physiological states for the better. Images used vary from study to study but frequently include general positive images like bright sunshine and specific images tailored to ailments. Someone who calls to mind an image of a tumor melting away in a candle flame is using a specific tailored image. Numerous health benefits have been reported. For instance, in a study on chronic abdominal pain in children[1], both the frequency and intensity of the painful episodes were found to be reduced following imagery.

To introduce the relevance of perception, consider simply that perceptual processes are involved in forming visual images of things that aren't actually there. Mental imagery has, in fact, been researched in the field of perception more than one would expect for such a tangential human ability. To fully understand why this ability was met with enthusiasm by researchers requires going too far afield into the history of psychology. The short version is that there was a time in psychology when all mental events were denied. Human behavior was seen as predictable just by the stimuli present in the world—a "real science"—that didn't have to be concerned with messy stuff that went on in the mind. Pavlovian conditioning mentioned earlier was a part of that behaviorist tradition. Perhaps you have also heard of instrumental conditioning's leader B.F. Skinner and his skinner boxes, rat mazes, and even his futuristic behaviorist novel, *Walden Two.* Skinner's name though gets mentioned a lot less than it used to. At one point, 90% of psychology departments' faculty and students were running animals in Skinner boxes or related equipment. Opponents embraced effects like mental imagery because they pointed to human abilities that could not be easily explained by a bleak stimulus-response behavioral analysis of humanity.

One type of imagery task that was studied inside and out involves *mental rotation.* Suppose you are given a drawing of a three-dimensional figure that looks sort of like what a deconstructed Rubik's cube might look like with arms made from little cubes that point in all different directions. Now you are given a second drawing which resembles the first but the cube-figure is oriented differently. You are asked if it is exactly the same figure as the first or whether it is instead a mirror image. To answer this annoying little unnatural question, you could try picking up one of the papers and flipping and rotating it to try to get the figure to be in the same orientation as the other. If one now would fit exactly on top of the other, it is the same figure. But if not allowed to lift the paper, many people can still get the answer correct. They rotate the figure in their mind. Now you can also see why anyone would use such an arbitrary insignificant task— I didn't go back to Roger Shepard and Jackie Metzler's groundbreaking paper on the topic[2] to see if they offer a justification, but presumably they wanted a task that could only be done through visual imagery, and complicated imagery at that. Indeed, there is individual difference in the ability to mentally rotate such polygons. A good deal of research was done with mental rotation specifically and visual imagery generally. Incidentally, Metzler switched careers shortly after getting her Ph.D. Perhaps preferring real to imagined stimuli, she became a veterinarian and was, in fact, my dog's vet for a time. Small world.

I will mention one more example of research involving imagery because it is relevant for perceptual learning. There is an irksome little perceptual learning phenomenon called the *McCollough Effect* that uses lines and colors. It will be considered briefly in the next chapter. For now, suffice it to say that the effect can also be obtained with lines and *imagined* colors. The bottom line of all the imagery research is that visual imagery in the mind can frequently substitute for real vision of things in the world.

Yet interestingly, despite the clear involvement of perception mechanisms in imagery, as well as the rich history of imagery research in the field of perception, guided imagery as a therapy did not derive from perception research. Instead, it has origins that include very different areas in psychology, such as Freudian psychotherapy and treatment for phobias.

The second common psychological intervention used for healing involves mindful meditation. Mindfulness meditation is an ancient spiritual practice in the Eastern part of the world and a component of Buddhism. Jon Kabat-Zinn's *mindfulness-based stress reduction* instruction at the University of Massachusetts Center for Mindfulness in Medicine, Health Care, and Society[3] has brought mindfulness meditation to numerous health professionals and patients in the United States. It is taught in hundreds of clinics, has been given to thousands of patients with conditions including heart disease, cancer, AIDS, and chronic pain, and is the subject of well over a hundred published scientific articles. But what is mindfulness meditation?

An exercise central to Kabat-Zinn's training program is known as *body scanning*, in which the participant focuses attention on successive parts of the body beginning with the left foot and ending with the head. The goal is to become mindful of the precise feelings and sensations that are occurring. Although not usually discussed in this way, consider that this body scanning technique can also be described as involving the body schema and proprioception. These are concepts which have been known in the field of perception for more than a hundred years and should also be familiar to readers of the present work. We have discussed the importance of keeping track of the positions of the parts of the body at all times, the ability to adjust its envelope through the addition of inanimate tools to the body schema, and the interesting way in which those tools are encoded neurally. Proprioception, or the felt position of parts of the body, is also critical in adaptation and can be readily modified when it drifts from accuracy, as seen in Chapter 9.

Other exercises in mindfulness meditation include paying close attention to breathing, to one's own thoughts, or to sounds "...just hearing what is here to be heard, moment by moment...just hearing them as pure sound" (pp. 73). Attention is clearly central to meditation. Attention, too, has been indispensible in perception. In the current context of perceptual learning, we saw that it is necessary to get the perceptual expertise type of perceptual learning. Attention must be paid to the orientation of a line, for example, to improve the ability to detect the difference between lines of similar orientations. Mindful meditation involves a greater than usual attention to exact perceptual stimulation occurring either within the body or in the world. Meditation already starts to feel de-mystified when re-described as an extraordinary ability to pay enhanced attention to the both the usual objects of perception and those items that ordinarily escape our attention. We have also seen that the process of perceptual expertise can turn ordinary perceptual abilities into extraordinary ones. The procedures of mindfulness meditation suggest a starring role for perception.

(And a parenthetical paragraph: I have already bored you enough with history but if you are so inclined, look up the perception movement known as *Introspectionism*. I suggest that there are connections to be found there between this classic approach in perception and paying unusual amounts of attention to exact perceptual stimuli that occurs in meditation.)

Step 3: Mind-Body Interaction is Perceptual Adaptation

Finally, and most importantly, I suggest that observed physical change to psychological interventions in mind-body medicine is not even just generally and vaguely perceptual but is specifically another example of perceptual adaptation, much like spatial adaptation to prism-displaced vision[4]. Prism adaptation is the classic example of the adaptation type of perceptual learning. For those who have not read the last few chapters, an observer looks at her own hand through a prism with the thick end to the right. She feels the hand to be in one spatial location but sees it in another a few inches to the right because of the prism. The perceptual system has a constraint that one object cannot be in more than one place at one time, including one's own hand. Consequently, if it is concluded that the visual and proprioceptive input refer to the same hand, then the discrepancy between the seen and felt positions of the hand indicates that there is something wrong with one's own perceptual machinery; adaptation is the process by which this inferred internal error is fixed. Either the felt position of the hand will shift to the right or the perceived visual location nudged to the left so as to remove the discrepancy.

J. Edwin Blalock is a neuroimmunologist who argued that the immune system is a sense organ. Blalock is also one of the early discoverers of shared messengers between the immune system and the nervous system. He says of the immune system: "A sixth sense, if you will, that completes our ability to be cognizant not only of the universe of things we can see, hear, taste, touch and smell but also the other universe of things we cannot. These would include bacteria, viruses, antigens, tumor cells and other agents that are too small to see or touch, make no noise, have no taste or odour"[5]. If the immune system is a sixth sensory modality then I suggest that a visual image of one's self fully healed along with an immune system that has detected something is wrong reflect a conflict between modalities, much like the conflict between vision and proprioception in prism adaptation. The conflict between the vision and immune modalities implies that one or the other must change to remove the error and bring the modalities back in line. If the immune system is the modality that changes then we say that a mind-body connection has occurred.

Figure 11-1 illustrates adaptation both for a vision-proprioception conflict in prism displacement and for the visual imagery-immune system conflict suggested for mind-body healing. You may recognize the left half of the figure from Chapter 9. I suggest that whether vision is placed in conflict with proprioception over the location of an object or with the immune system over the identity of an object, the form of the result will be the same. Before adaptation, there is a conflict between modalities that must be addressed. Consequently, during the process of adaptation, at least one of the modalities must shift. When the process of adaptation is complete, successful adaptation realigns the two modalities such that they are once again in agreement.

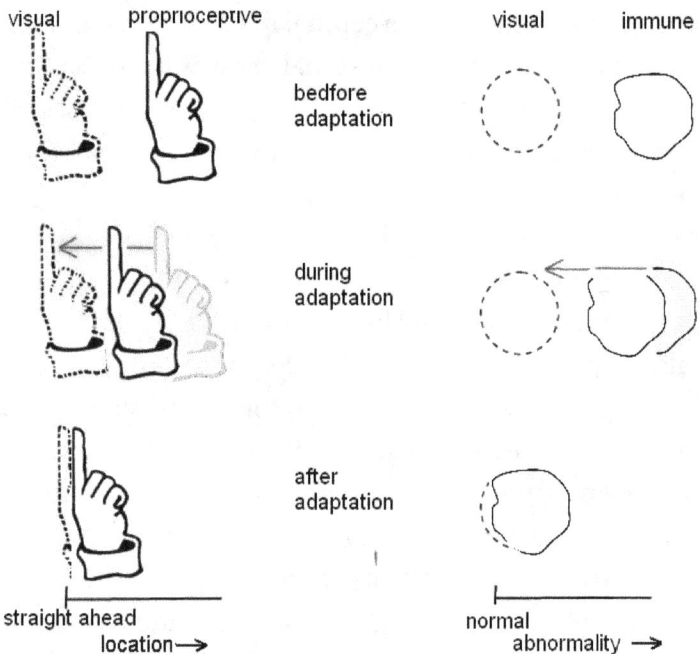

Figure 11-1 Adaptation to a conflict between vision and proprioception concerning location of an object (left) and between visual imagery and the immune system concerning the health of a cell (right). Top: before adaptation when there is a conflict between the information obtained by two modalities. Middle: what happens during adaptation when one or more modalities will shift in the direction of the other. Bottom: after adaptation when recalibration is complete and the modalities once again provide identical information on the shared parameter. From Bedford, F.L. (2012). A perception theory in mind-body medicine: guided visual imagery and mindful meditation as cross-modal adaptation. *Psychonomic Bulletin and Review, 19,* 24-45.

Experiments that use guided imagery for healing often include instruction for images that are, or will lead to, being fully normal, healed, restored and give a sense of well being. This kind of visual imagery provides the conflict with the immune system information. Consistent with the view that mind-body interaction reflects cross-modal adaption is the finding that participants who report the most vivid visual images following the very first training session also show the greatest healing. Psychologist Nicholas Spanos and colleagues showed this in a series of experiments on warts[6]. Visual imagery ability varies widely in the population and in the present theory, we would expect that the greater the imagery ability, the greater the cross-modal conflict and hence change. I also found this in a study on pain and assorted skin conditions (see later) and additionally found that participants found it harder to "visualize your condition healed" compared to "able to form vivid visual images during the technique". It may be especially difficult to visualize a counterfactual, the very ingredient needed for the adaptation/healing to occur. If this difficulty occurs, it may be the immune system that is changing vision rather than the other way around.

The theory may explain why some conditions appear to be especially amenable to change through psychological intervention. A patient with the potentially fatal disorder dermatomyositis was cured following the better part of a year with visual imagery and (transcendental) meditation[6.] Dermatomyositis is a potentially fatal autoimmune reaction that affects muscle tissue and skin; the inflammation and degeneration lead to debilitating muscle weakness and skin rash. The physician authors (Collins and Dunn) attributed the benefit to changes in the humoral immune system. In the present theory, autoimmune disorders should be especially changeable by mind-body interactions because they are perfectly suited to cross-modal adaptation. In autoimmune reactions, the body launches an attack on its own cells, mistaking them for foreign invaders. That is, *the percep-*

tual system has made an error! It is therefore an ideal candidate for correction through adaptation. The very function of adaptation is to correct internal errors in sensory systems, whether they be in vision, proprioception—or the immune system.

Numerous reports in the medical literature on successful psychological interventions for skin disorders may also be sensible in the present view. Note first that folk wisdom also singles out skin conditions as especially changeable. In self-help books, it is noted that magical thinking and power of suggestion has been used for years by pediatricians to cure warts in children[8]. Responsiveness in chronic skin disorders may in part be due to errors made by the immune system. For example, warts result from the immune system being tricked by the papilloma virus into providing "free room and board"[8] and psoriasis is believed to involve an autoimmune component. Two additional factors may contribute to the special effectiveness of imagery on skin conditions. First, we get to see our own skin frequently. This should make visual images of skin in the mind's eye relatively easy compared to conjuring ordinarily invisible internal organs like kidneys. Second, the skin itself is involved in the sensory modality of touch, unlike internal organs or tumor cells. Interactions that may occur when multiple modalities are present— vision, touch, and the immune system—warrant further investigation.

Faith Healers and Adaptation

Excitement and wailing can be heard in the background. There is a laying on of the hands, a question about faith, a declaration of healing—and soon Johnny is no longer in excruciating pain. It's a miracle!" someone shouts. And it is. Until Johnny is found dead two days later from a ruptured appendix. This type of example has led skeptics to the conclusion that mind-body interaction only makes people think they are healed but does not affect the disease process itself. Such seeming failures are readily explicable within the perceptual adaptation theory of mind-body effects. Other more recognized instances of perceptual adaptation also have the equivalent of faith

healing. In prism adaptation, you may recall the disagreement between vision and proprioception is often resolved by shifting the
proprioceptive (felt) location of the arm to agree with vision. This
occurs even though it is the visual location of the arm that is incorrect because of the prism. Hans Wallach, another long-time prolific
perceptual hero, used the label *counter*adaptation (italics added) to
describe the effect. He used the label to draw attention to the fact
that such a change to the wrong modality is anything but adaptive. If
not in the confines of a protected laboratory, mistaking the error in
visual location for an error in arm location would be as harmful to
survival as mistaking the hot appendix for a minor ache. The faith
healer did cure the pain but curing the pain in this instance was a bad
idea. The pain was right, the imagery wrong.

When I was an undergraduate at the University of Pennsylvania
and already doing my own research on adaptation, I think my advi-
sors no longer knew what to do with me and so shipped me off to
meet Hans Wallach. I took the suburban train from 30th St. Station in
Philadelphia to Swarthmore college where he met me outside. It was
an unusually windy day and my very long hair was positively *flying* in
the wind. Perhaps he thought the rest of me was soon to follow. In
any event, Hans Wallach found it very amusing and hair-in-the-wind
seemed the only thing he wanted to attend to. I was mortified. I
thought he would not take my research seriously and I was one very
serious student. I was wrong, though, and we had an unforgettable
conversation. I learned much later that a sense of humor is, of course,
a sign of great intelligence. He passed in 1998 at age 93. Hans Wallach advanced our understanding of perception and perceptual
adaptation with more than a half a century of brilliant research. I can
still hear his laughter.

Returning to the main event, seeming failures in healing can also
be understood with adaptation. Sometimes, the results of imagery on
physical aspects of cancer have been disappointing. For instance,
women receiving *relaxation visualization therapy* (RVT) while
undergoing radiation treatment for breast cancer[9] did not show

biological improvement as measured by T cells and cortisol levels, though did show reductions in depression, stress and anxiety. Like the anecdote with the pain, failure to show biological improvement may result from a resolution between the conflicting modalities that does not suit the host. A conflict between vision (normalcy, well being) and the immune system (something wrong, under attack) can lead to adaptation in which vision wins and the immune system no longer detects a threat. Getting the immune system to stop attacking is what is needed in autoimmune illnesses but would be exactly the *wrong* resolution in cancer. What one wants instead is to turn *on* the immune system in the event the immune system had failed to detect the cancer cells. This would require conflicting visual information that something is very wrong, not that it is healed or peaceful. RVT included visualizing the breast fully healed and imagining tumor cells under attack by the immune system. "Fully healed" imagery would provide the wrong conflict for cancer, visualizing tumor cells the right conflict, and visualizing an attack likely the wrong conflict. This type of mixture is fairly common in imagery training for cancer and also one of the most common alternative treatments for cancer.

Guided imagery treatment includes images that emphasize healing, positivity, and inner strength, along with seeing the tumor cells in the mind's eye. It may be counterintuitive for practitioners to use imagery that emphasizes illness, which may account for some failures of imagery's effectiveness on biological markers in cancer. But imagery that emphasizes illness is exactly what it is needed for cancer. Note also that cancer treatments currently have fewer side effects than they have had previously. It would be interesting to compare equivalent chemotherapeutic agents under conditions in which there are visual makers of illness, such as hair loss, and when there are not. The present theory predicts the former should be more effective than the latter, despite our natural repulsion to such treatments.

In general, whether mind-body interactions work and whether the outcome is adaptive are two distinct issues.

Why Mind Body-Interactions Occur

The theory may also help explain why mind-body effects exist at all. In psychology and biology, we have come to expect that tasks most important for our future well-being remain protected from cognitive abilities of knowledge, expectations, beliefs, wishes, and conscious reasoning. Such cognitive processes tend to be slow, subject to false beliefs, inconsistent, and require attention, any of which could disrupt essential functioning. For instance, sexual attraction is not determined by rational cognitive reasoning, as everyone laments at one time or other. It is determined instead by evolutionarily defined high mate value. We may be able to literally sniff out and pick the most dissimilar immune systems, according to the results of research, but it just feels like attraction. Vision is often cognitively impenetrable as well. (Try again the demonstration with the Müller-Lyer visual illusion from Chapter 5.) Blood pressure, heart-rate, remaining upright, and maintaining the right acid-base balance in the blood are typically also outside the influence of our mental whims. One would think certainly then that the immune system would also be an example. Why *should* what we think influence such an important job? I once asked Andrew Weil this question following a colloquium at the University of Arizona and he did not know.

In the present theory, the mind-body interactions do not reflect a higher-order cognitive influence on the body at all. This is why they can occur. Instead, their evolutionary function involves the interaction between perceptual modalities, just as vision and audition or vision and touch naturally interact. Interactions between systems that address the same parameter serve an important adaptive function of checking for errors in perception. Otherwise, there would be a problem knowing when a perceptual output resulted from an error or when it was indicating something new about the world (see also Chapter 12). We saw this, too, when evaluating different possible reactions to a disagreement between modalities (Chapters 8 and 9).

Mind-body interactions only appear to be mind over matter.

The Perceptual Sixth Sense of the Immune System

The cross-modal adaptation account of healing depends on neuroimmunologist Blalock's ingenious assertion that the immune system is a sense modality, like vision or audition. How should the field of perception respond to this claim?

Like other sense modalities, the purpose of the immune system involves the detection of entities in the environment. The comparison is off to a good start. The immune system has a lot of parts, such as innate immunity, acquired immunity, humoral immunity, and cellular immunity. It may seem like the immune system consists of too many separated components to be considered a proper single organ of perception. We can exclude from consideration for perception the parts of the evolutionarily older innate immunity that are mechanical (e.g., coughing) and chemical barriers (e.g., enzymes in saliva)because they do not function as detectors even if they do protect against foreign invaders. That still leaves multiple systems: the parts of innate immunity that do detect foreign cells but do not learn, acquired immunity which learns to protect the host against repeat offenders and its further subsystems of humoral and cellular immunity. Humoral immunity (b-cells) is comprised of cells that come from bone marrow, detect unprocessed antigens, produce antibodies and complement to destroy foreign invaders, and are especially good at foreign objects whereas cellular immunity (t-cells) comes from cells that mature in the thymus, detect peptides, are directly destructive, and especially good against cells of the host that have been infected.

Traditional sense modalities are also comprised of multiple parts. Vision consists of multiple systems, with distinct rods and cones at the receptor level that are optimized for different lighting conditions, distinct ventral and medial streams higher up, and an evolutionarily older vision system mediated by the superior colliculus rather than the cortex. There is both a primary olfaction system and an older accessory olfactory system believed involved in pheromone detection. The modality summarized as *touch* or the *cutaneous system* in perception textbooks consist of the very distinct properties of haptics (recognition of spatially extended objects through active exploration, as in the experiment from Chapter 4), proprioception (localizing the position of body parts as discussed in previous chapters), skin sensations of temperature, touch, vibration, and texture, and the perception of pain which we will consider shortly. Multiple systems clearly have not been an exclusion criterion for a sense modality.

The spatial distribution of the sensory receptors in modalities can also be considered. The receptors for hearing all coalesce in a single region in a discrete organ of the ear. Likewise for vision, taste, and smell, within, respectively, the eyes, the tongue and the nose. The immune system, on the other hand, has receptors distributed widely throughout the body. Too dissimilar? Not really because the immune system is similar to parts of the modality of touch, in which receptors for temperature, light touch, and heavy pressure are distributed throughout the body. In addition, human vision evolved from light sensitive spots that were widely distributed, a sensory system that is found today in the earthworm. Thus, a particular distribution of receptors is not a criterion for a sensory system. Nor is the number of receptors. The humoral immune system is capable of producing at least 100 billion distinct receptors for new chemicals which seems qualitatively different than human vision's three fixed cone receptors on the retina. However, "striking similarities" of the immune system to olfaction have been noted[10], with olfaction also able to detect a practically infinite number of new distinct chemicals.

It can be argued that perception detects entities in the external environment but the immune system is confined to those inside the body. While an inside-outside distinction seems like a qualitative difference, closer analysis suggests more of a continuum of range of operation for sense modalities. I suggest that the senses can be roughly ordered from near to far detection ability: immune system, taste, touch, haptics, olfaction, audition, and vision. The gustatory system provides detection at a boundary between inside and outside the body, with taste immediately preceding the introduction of substances into the body and with receptors found down into the throat as well as on the tongue. Touch operates close by at the surface of the body and next, objects can be discerned through haptics a little further, within arms' reach. Finally, the remaining modalities are more distal senses and can gather their information from quite far away. *Rather than being singled out from other sense modalities, the immune system seems to complete the range of detection.* The immune system operates closest to our core, picking up where taste leaves off. All the sense modalities together ensure full coverage for any distance where an entity may be lurking. See Figure 11-2 (opposite start of chapter) for an illustration.

Finally, I am reminded of a famous question that played a central role in my prelim papers many years ago: Why do things look as they do? By extension for audition, why do things sound as they do? What would be said of the immune system? Why do things __ as they do? The closest word we have is "feel", but it would not be quite accurate. Perceivers are familiar with the outputs of sensory modalities being accessible to conscious awareness, description, and deliberation. We are aware that we hear the baby, smell the skunk, feel his hand. We may be aware that we have a fever or lethargy (*sickness behavior*), but do not "perceive" the bacteria or cancer cell that caused them in the sense of conscious awareness of what the sensory system has detected. This is a notable difference, but not a deal breaker. This is because the output of the traditional modalities can be dissociated from conscious awareness, even if conscious experience frequently accompanies them. Remember the disorder blindsight from Chapter 1? Patients see, but don't know they can see. Blindsight is one example of a dissociation in the visual modality between seeing and having a conscious perceptual experience. Within touch, a hot stove causes the hand to withdraw faster than the experience of pain or anything else—that is, we have some kind of non-conscious perception of dangerous heat before the conscious perception of the heat. Another example of perception without conscious perception comes from olfaction. It has been reported that children can detect a biological sibling through smell which may adaptively prevent sexual attraction to close blood relations. Yet I claim no one is known to say, "I'm not into him because he smells like my brother."

Perhaps the output of immune system detection is not available to conscious experience because it is a very old system, evolved before conscious experience were possible. Perhaps it is not available to consciousness because there was not any adaptive benefit to having the detection of a cancer cell available to higher-order cognitive processing (taking the next flight to Sloan Kettering is too new to count.) Whatever the explanation for the absence of conscious experience of immune system output, the dissociability of perceptual outputs from consciousness in other modalities implies that it should not exclude the immune system from the list of perceptual modalities. The differing role of consciousness in the immune system compared to other sense modalities may also contribute to unique aspects of the interaction between meditation/imagery/vision and the immune system. (See section below, "body scan and mindful meditation".)

In conclusion, the immune system remains a candidate for a perceptual modality with similarities to other more traditionally recognized sensory organs, such as vision and olfaction. There are similarities in function (detection), input (specialized receptors), internal representations, and even output (need not result in a conscious experience)[11].

Oh My Aching Head

Mind-body interventions are quite successful with pain. This deserves special consideration. Is the "subjectivity" of pain a simpler and more fitting explanation for why mind-based interventions work than is an immune-visual perceptual adaptation theory? It may appear as if the immune system is not relevant for pain even if it is relevant for autoimmune disorders, skin conditions, and cancer.

However, some of the earliest brain messengers that were shown to be made in the immune system and not just in the head were endorphins, which modulate pain. In addition, the receptors for opiates, which are involved in pain sensitivity, remain one of the best studied classes of neural receptors that are also found in the immune system. Later, macrophages in the immune system were shown to be a source of substance P, also a brain chemical involved in pain. Currently, many levels of interaction between the immune system and pain thresholds in acute and chronic pain are recognized. The immune system can clearly be involved in the regulation of pain.

It may also seem like visual imagery of pain just is not sensible. Imagery in pain is certainly not as straightforward as imagery in skin disorders. One can see, either literally or in the mind's eye, skin as red or normally colored, as scaly or smooth, as bumpy or flat. It is intriguing to contemplate precisely what visual images are involved with pain. Does a person form an image of himself hunched over? Or hand on her aching head? A straight arm visualized as unable to bend? Imagery instructions sometimes have included relatively specific instructions, such as to imagine warm sunshine soaking into painful areas, but often are unrelated or general, such as to visualize a pleasant scene, to imagine being healthy and strong, to imagine anything they went, and to emerge with a sense of well-being. Studies typically do not investigate what subjects actually do after receiving instructions. With a different cognitive ability, specifically blindfold chess playing, psychologist and chess master Eliot Hurst informs me that the imagery becomes more abstract as people become experts. Although chess playing is quite distant from experiencing pain, it raises the possibility that abstract images effective on pain are being developed with the extended practice that mindfulness training entails. Indeed, one advantage of imagery would be its ability to conjure images not otherwise possible with real vision. To the extent that "perception of pain" is itself a modality, the existence of cross-modal conflicts between vision and pain in addition to those of vision and immune system may also contribute to the particularly effective pain control reported with mind-body interventions.

Attributing successes instead to the subjectivity of pain is not an actual theory nor is it even very accurate. Pain has seemed different than other perceptual experiences because one person cannot verify what another person claims. In vision, a claim that one line looks longer than another in a picture can be checked by looking at the same picture and seeing if it looks longer to you, too. In pain, you cannot reproduce the exact stimulus causing the pain and use it to check if it has the same effect on you—nor would you want to if you could. It gets thought of then as "subjective". Labeling pain as subjective is an inappropriate dismissal because pain is mediated by chemicals and neural responses every bit as real as the chemicals and neural responses that mediate other physical and perceptual manifestations. The casual view that pain is all in one's head may owe less to science and more to the observation that pain is of no consequence except to the sufferer.

Mindful Meditation, Self vs. Not Self, Consciousness

The preceding discussions have emphasized visual imagery. Meditation also plays a significant role in both healing and perception. Mindful meditation calls for what can be described as an unnatural, or at least atypical, hyper vigilant attention to the object of meditation, such as the body or sounds or the breath or thoughts. "If it moves off the breath hundred times, then you just calmly bring it back a hundred times, as soon as you are aware of not being on the breath"[3]. In prism adaptation, recall that when two modalities are in conflict, not attending to one of them makes it more likely to change (the *directed attention hypothesis* in Redding's older version of the effect he discovered). This seems to be a rather maladaptive conflict resolution. The modality that should change is the one determined to be at fault, not the one that happens to escape attention. Nonetheless, there is elegant evidence that it occurs in prism adaptation.

Consequently, for vision-immune system conflicts, attention to one modality should also push the other to change. Meditation can provide precisely that attention. The extra attention to the visual imagery that occurs when also simultaneously meditating should lead to the immune system being especially likely to change. In addition, the single-minded focus of meditation means that only the intended images will get through. Meditation should prevent the intrusion of inappropriate imagery that would dilute, and perhaps even undermine, the changes to the immune system that one is trying to bring about. The unusual attention properties of meditation therefore also allow for concentrated training that should facilitate adaptation, for vision-immune or vision-proprioception conflicts.

Imagery by itself then, without meditation, should be sufficient to bring about healing, provided that appropriate vivid images can be manufactured in the absence of special attention assistance. Moreover, the combination of imagery and mindful meditation should be especially effective. This prediction of the present adaptation theory is in contrast to Kabat-Zinn, whose position on the relation between imagery and mindfulness is as follows: "In order to be effective for healing, we believe that the use of visualization and imagery needs to be embedded in a larger context, one that understands and honors non-doing and non-striving"[3]. Psychological interventions typically use multiple techniques together making it difficult to assess the separate contributions of each component.

When perceiving the world, attention serves to bring particular things into conscious awareness. Unlike the traditional sense modalities, conscious experience of the immune system is limited to general sickness behavior, like lethargy, as noted earlier. It is intriguing to think that meditation, with its seemingly super human attention powers, may also boost the ability to become conscious of health-related things that we normally are not.

The body-scan exercise of mindfulness meditation hones awareness of the body schema. Performing the body scan exercise may be reminding the mind, over and over, where the current boundaries of the body are—attending to where we begin and where we end, distinguishing self from non-self. There is a notable parallel in the immune system. For half a century in the field of immunology, a central role of the immune system has been considered to be its ability to distinguish self from non-self. Without this ability, the immune system would not know what is safe and what to attack. In both traditional perception and the immune system, representation of the body boundary is changeable. In body schema, "self" can be added to with inanimate tools or subtracted from, such as the rewiring that occurs following loss of a leg or stitching two fingers together. For the immune system, autoimmune illness reflects a failure to tolerate the self, a very literal meaning to being one's own worst enemy. The body attacks specific parts of the self, "believing" they are, in fact, not part of the self. The full potential of this parallel has yet to be realized.

A Call to Action

There are many consequences of the theory to explore. Some of these may even have practical benefit. Real vision—not just vision through mental imagery—should also lead to healing. The trick would be how to create a real stimulus that is appropriate, not just an imagined one that has the advantage of being conjured up at will. For skin conditions affecting one arm, reflecting the unaffected limb in a mirror would give the appearance of two healthy arms. Mirror therapy has been used in other contexts and may work here to create a visual-immune system conflict even more powerful than imagined vision. A second pursuit is to try the counterintuitive approach that negative-type imagery may be more effective for cancer than more typical positive imagery instructions have been to date. By negative imagery, I mean those in which one sees ugly tumors rather pretty meadows. And treatments with plenty of visible side effects. Third, if guided visualization and meditation do not appear to be effective, check to see if there was a change in vision rather than the immune system. If so, try to shift the conflict resolution back to a change in the immune system, perhaps through increased attention to vision. Finally, giving simple tests of imagery may allow people who would most benefit from these perception interventions—note I have switched from calling them psychological interventions to calling them perception interventions—to be easily identified. Those who do not test as vivid imagers can be given training to improve the skill first before spending what would otherwise be wasted time on mindfulness-imagery treatment.

But always the theorist, I cannot help but point out a few of the more abstract questions that the theory raises. Some of those, too, may eventually lead to positive benefit, though not all will. Ethical considerations aside, can imagery be used to make health worse as well as better? What does imagery for pain look like and does it change over time? Are mind-body interactions best used in instances where the immune system has made an error, such as in autoimmune diseases like rheumatoid arthritis or Sjogren's syndrome? Can the body scan exercise be used in autoimmune illnesses to inform where self begins and ends? Can we uncover examples where the immune system cooperates, rather than conflicts, with vision, hearing, touch, olfaction, and taste? Would it be fruitful to think of training attention through meditation as enhancing the machinery itself that in turn usually enhances perception in the perceptual expertise type of perceptual learning? Finally, will unnaturally focused and sustained attention also help produce effective change not just in the immune system but in our two types of perceptual learning more generally, expertise and adaptation?

Imagine Your Pain Away

Meditation training can be a big time commitment. Buddhist monks undergo years of training to reach a state in which accomplishments like mentally changing body temperature are possible. I wanted to determine if a very fast version of a combined imagery and meditation intervention would be of benefit for physical healing. I was also interested in whether individual differences in imagery ability might influence the effectiveness of imagery, as Spanos and colleagues had found. Finally, I wanted to use photo documentation wherever possible to develop an objective measure of any healing that may have taken place.

Consequently we created a short 12-minute audio file for the training or treatment—depending on how you view it—consisting of half body scan exercise and half imagery[12]. The body scan was adapted from Jon Kabat-Zinn's mindfulness meditation program, except that it lasted less than 6 minutes rather than 45 minutes and was performed sitting up. Participant's attention was directed to the left foot followed by successive parts of the body up to the head. The imagery half included mentally going to a favorite place, experiencing the surroundings, looking at the affected part through a magnifying glass, seeing the ailment under attack, and visualizing full healing. Following an initial session in the lab, we instructed the subjects to practice at home with or without the audio file and document on a log over a 3-week period at least 8 more times for at least 10 minutes each time. This totaled to only about 90 minutes of accumulated training. To assess imagery without lengthening the experiment, we asked at the end of the very first training session in the lab simply for a rating of the vividness of the preceding images. The rating scale ranged from 1 ("couldn't see anything") to 10 ("like it was really there").

We recruited 47 college students with chronic ailments for the experiment. Since college students are typically a healthy bunch, participants were invited based on a questionnaire administered to a large number of undergraduates. The questions were designed to screen for chronic conditions along with an interest in a non-invasive training intervention. Table 1 shows the responses to the questions. Despite young adults' excellent health, note the high incidence of reported Type A personality, a known risk factor for later health problems. Also note the large male-female difference in incidence of headaches, the most frequent specific ailment in women. The two largest subgroups in the experiment were people with 1) skin disorders characterized as dermatitis, warts, acne, psoriasis, or unspecified rash each involving the hands, arms, and/or face and 2) pain, including headache (migraine and tension type), injury-related, arthritis, endometriosis, herniated disk and unknown, affecting head, arm, knee, neck, back, pelvis, or abdomen. The duration of their conditions ranged from 3 weeks to 18 years with a mean of 6.8 years so some of these people had stubborn cases that clearly had been resistant to conventional treatment.

Table 11-1 Percent "True" Responses on Health Screen Questions

	Female	Male
number of subjects	604	385
Skin problem, hands arms	9.4	5.5
Skin, face	22.2	26.8
Headache, frequent	39.1	15.6
Pain, other, chronic	8.8	7.0
Autoimmune	1.8	0.8
Sinus, frequent	19.7	17.7
Chronic, any	17.7	16.1
Interest in Training	40.2	38.7
Type B personality	65.6	70.4
Type A personality	46.2	39.5

Questions given to 1008 undergraduates at the University of Arizona. Type A and Type B do not sum to 100% because they were asked as 2 separate questions.

Participants were assigned to either an experimental group that received the intervention or a control group that did not. Control group participants were told that they were on a wait-list and could return in 3 weeks. All participants were asked 14 questions, including to rate the severity of symptoms right now. A photo was taken of skin conditions. All participants returned in 3 weeks to retake the questionnaire, answer a question concerning general imagery ability, and have another photo taken. Training-specific questions were additionally asked of the experimental group participants.

Among those who returned three weeks later (all but 5, comprised of 3 from the experimental group and 2 from control group, all of whom reported being too busy with end of semester burdens), nearly all participants in the experimental group responded favorably to what was being asked of them. That is, they thought others would be able to engage in the training and gave relatively high ratings on a 10-point scale for ability to follow the instructions and get themselves to practice. Remarkably, they also almost all reported at least a small benefit on symptoms. Interestingly, they rated the benefit of the treatment for other things, usually stress, higher than the benefit to specific symptoms they were working on. Also quite notable is that lower ratings were given to being able to specifically visualize the ailment completely healed compared to being able to form other vivid images during the technique. Table 2 shows these results.

Table 11-2 Results of Imagery & Mindfulness Training

	Questions	Results
1	Think others will be able to engage in the technique - Percent of people answering "yes"	95.8%
2	Able to follow instructions of technique - Mean rating on scale from 1 to 10	8.0
3	How easy to get yourself to perform the technique	7.2
4	Any benefit for condition/symptom you've been working on (Percent of people with responses of "2" or higher on rating scale)	4.8 (92%)
5	Any benefit from technique for anything else (e.g. stress)	8.0
6	How easy to visualize the ailment as fully healed	5.8
7	How well could you form vivid visual images during the technique	7.9

The differences between questions 4 and 5 and between 6 and 7 are statistically significant; t(23)=7.54, p<.001 and t(23)=4.51, p<.001, respectively.

Results involving individual differences in imagery ability are shown in Figure 11-3. There was a substantial relation between the visual imagery vividness rating at the end of the first training session and the change in self-report of symptoms from pretest to posttest. Change in symptom rating (rating of how bad symptoms are right now on a scale from 1 to 25; 1 = "completely normal", 25 = "as bad as I've ever had it") is a semi-objective measure despite being a self-reported measure because of the delay between tests and the presence of other questions. That is, posttest response is largely independent of pretest response because it is unlikely subjects remembered the exact score they gave previously. This means we can conclude

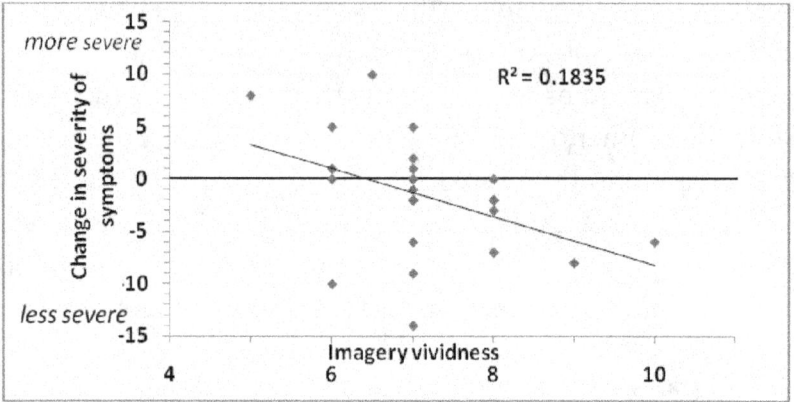

Figure 11-3 The higher the subject's imagery ability, the greater the improvement from the imagery-meditation intervention (less severe).

Participants with the most vivid visual imagery showed the greatest improvement in self-assessed severity of symptoms. The correlation appeared both in women (-.42, n=16) and men (-.52, n=8) and for both skin conditions (-.62, n=10) and pain (-.46, n=9). Consequently, the relation between imagery and improvement is not also an artifact of averaging men and women, of averaging outliers (see Figure 11-3) or of participants rating everything highly (not shown).

People who image vividly do better with the imagery-meditation training. Interestingly, in the control group, people who were vivid imagers did worse! Perhaps spontaneous images without training have been helping sustain chronic conditions in those who are high imagers. This may be an example of imagery making things worse rather than better. If conflicting images can make symptoms better than we would expect that consistent images should make things worse. In general, there was a trend for women to show more improvement than men, but it did not reach statistical significance and larger numbers of subjects are needed to explore any gender differences. Figure 11-4 shows photos of one subject's skin condition before and after training.

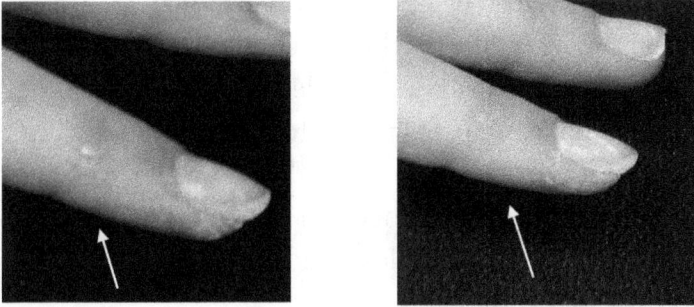

11-4 Photo of a subject's finger before (left) and after only three weeks of the imagery-meditation intervention. Compare the white spot (wart) in the 2 pictures.

The experiment was successful in implementing a streamlined paradigm for mind-body healing that ordinary busy people can engage in and find satisfying. The vast majority of participants felt the training was of some benefit, which is impressive given frequent dissatisfaction with conventional treatments for chronic conditions. The experiment also found that people able to form vivid visual images during training were already able to have self-assessed reduction in severity of symptoms despite only about 1 and 1/2 hours of cumulative practice. I especially like the finding that people had such a hard time imaging the condition fully healed. This is consistent with the theory that the imagery sets up a conflict. It indeed may be hard to form an image of something that is the opposite of what is true. This is something to work on—to get past this resistance—to produce the very image that is needed to get change.

Further research, both at behavioral and immune system levels, will continue to demystify mind-body connections and lead to additional effective treatments. These connections are neither myth nor supernatural and hopefully they will continue to be embraced by the community. I hope in particular that my perception colleagues at large will do so.

Adaptation is a powerful mechanism. Different senses disagree about the same object or event. The disagreement indicates that there is an internal error and the sense modality that made the mistake is fixed. This ensures that perception can be accurate in the future. Healing, immune system, guided imagery, and meditation are very exciting applications for perceptual adaptation. I currently do not have a good name for the present theory and have been referring to it as Perceptual Learning Theory–i (for immune system). I invite others to meditate on a suitable replacement.

12

The Finale and The Future

Time is measured in heart-throbs[1]. An object you control is seen to move before you actually move it. Your arm feels to be three inches to the right of where it was just minutes ago, though it hasn't actually moved at all. A tennis racquet feels like it is now part of your body. A Native American man is finally recognized correctly after many misidentifications. You can identify all of the ingredients in an Italian dish by taste following cooking lesions. Perception can be changed.

We have seen a lot of change. It is time for a break. For me, that means a theory break. Yes, you heard right. I like theory, the more abstract the better. Those who do not share my proclivities can take a different break—maybe try the tasks in the next and last aftermath chapter and then jump back to the Future in this chapter. And see Figure 12-1.

Figure 12-1 No one knows if you are thinking deep theory.
Marcelo Rampazzo /toonpool.com".
Under license. Reprinted with permission.

FINALE

Putting It All Together (Perceptual Learning Theory)

The tree of learning that was started at the very beginning of the journey (Figure 1-1) can now fleshed out a little more. At the top of the tree is experience (Figure 12-2, facing end of the chapter). Only when experience "makes us better" are the effects of experience to be considered learning. That was discussed in the first chapter and we can pick up from there. If one branch of the tree is learning, I am tempted to refer to another branch as *not learning*. However, if I did, my colleagues would accuse me of committing my very own *Not-the-Liver Fallacy*. I devised the Not-the-Liver Fallacy to point out that psychologists often turn one discovery into two by assuming that what's left after their discovery is also a discovery. If you were the one that found the organ responsible for a few bodily processes (the liver), would you also conclude that everything the organ does not do is also an organ (not-the-liver)? Of course not. It is a particular problem in neuropsychology, an area where researchers try to infer psychological processes from damage to regions of the brain, but the problem occurs throughout psychology.

I will try not to make the same mistake here. I will, therefore, tentatively refer to branches other than learning as injury, comprised of things which make us worse, and fatigue, which may or may not make us worse. If fatigue does not make us worse then it may belong somewhere else on the tree. Hopefully, these more specific categories have more basis in something real than a tempting and convenient not-learning or not-the-liver construct. Choosing the right branches in that half of the tree, however, does not concern us too much here since our concern is with experience that makes us better—learning—rather than the processes that do not. One note first on the "worse" half of the tree before moving on. You can see where perceptual disorders may be located (Figure 12-2, right branch)—those cousins to perceptual learning discussed in the first chapter.

Returning to experiences that make us better, they can be subdivided into two major sub branches: processes that are about the external world and processes that are about the internal world. The former allow us to acquire new information about the world, whereas the latter are about correcting internal malfunctions. It is the world information processes that are traditionally regarded as *learning.*

This branch includes Pavlovian conditioning, a type of primitive associative learning in which we learn to predict one thing from another without needing to be consciously aware of the relation. You can see conditioning's influence by observing behavior. For example, you see a dog salivate to a tone in expectation of food that had been previously followed the sound of a tone or you see a person's eyes widen at the sight of a dog in fear of getting bitten again. You may witness a rat in the laboratory suddenly freezing in place when it hears the tone that predicts shock and a person might even automatically kick someone in the presence of a bright square if the square had repeatedly been paired with tapping his knee with a hammer. Also within this branch, memory is hard to separate from learning. You note that Uncle Maury only calls on Thanksgiving, but Uncle Sammy is available at least once a month. You remember this and store it away in case you need to recall the information on their birthdays. Or if you need to know the capital of Turkey, you can learn it from an electronic encyclopedia and hopefully remember it later. (If you have come across the term "implicit memory", you need not be concerned with this alleged type of memory because it was a result of the Not-the-Liver Fallacy!) Visit Seattle, Washington and Tuscaloosca, Alabama and you will learn about each place. What all of these things have in common is that they represent something about the world. They teach you something about the way the world works that you did not know before the learning.

Such world learning can affect you in a variety of different ways. You can be smug about your new knowledge of country capitals. It can change your whole belief system from trusting to cynical, for example. It can also change how you act on the world—only fish for salmon in the Northwest, not in the Southeast—and it can change who you feel you like (Sammy!), don't like (not so sure about Maury), and what you are afraid of. It can even change what sets off your most primitive unconscious physical reflexes. But what world learning cannot do is change your perception. Not in the sense of what you *literally* see, hear, smell, taste, touch, and *immunate* (new word coined here for our hidden immune system sense—see Chapter 11)

Contrast that with the sub branch of the experience tree that is about the internal world: It does change perception. Although this perception type of learning is less familiar to many learning and cognitive researchers and, moreover, may go unnoticed in everyday life, this, too, should be properly regarded as learning. That is because perceptual learning also leads to improvement from experience (See Chapter 1). It leaves us better equipped to survive in the world. World learning does so through the representation of environmental properties. Perceptual learning does so through improvements in ourselves. This in turn allows us able to apprehend the world correctly. The perceptual learning processes keep the sensory systems in good working order to enable accurate world learning. Traditional theories of "learning" (i.e. restricted to world learning) assume that perception is accurate and begin with that as a starting point. As we have seen, it takes a lot of work to keep perception accurate. The two types of learning must also be occurring in parallel, not perception first and world learning after.

Odd terms, like *plasticity*, sprung up to refer to perceptual learning. If plasticity is supposed to convey that perception is primarily fixed—like a rigid object—but can be changed just a little like a piece of plastic that has a bit of give before it sets, then I disagree. Perception does not just change a tiny bit once in a blue moon. Mostly, the term "plasticity" just confuses people and no one knows quite what it means. This may have been the intent since no one knew quite what to do with perceptual learning before. Now, it can take its rightful place within the grander schema of things as Figure 12-2, and hopefully this entire book, illustrates.

For both world learning and perceptual learning, there is error correction. Even for world learning, the motivation to learn is a difference between what you expect and what you get. That difference means you have made an error. But the difference is about which errors you are correcting: those concerning facts about the world—or those about yourself.

Returning to the tree, there is also likely a third major branch of learning. World learning involves matching your internal states to the world while perceptual learning often involves matching your internal states to one another. The third branch involves matching your internal states to the internal states of other people. Language is a great example. According to the famous linguist and political activist Noam Chomsky, languages are not out in the world. We can add that their acquisition is also not a matter of correcting sensory malfunctions. Instead what matters is that your language matches someone else's. Consider also driving. The only significance of learning to drive on the right side of the road is that other people in the United States drive on the right side of the road. Duplicating the same behavior in England would clearly cause accidents. Some of motor skill learning belongs in this category as well, despite a seeming resemblance to perception-type changes. Motor skills are not things that exist out in the world to be acquired nor is there any internal problem that gets fixed with their acquisition. Instead, you try to match what someone else can do. A ballet dancer tries to duplicate an exact form that others have done before her. Incidentally, there are neurons in the brain given the intriguing label *mirror neurons* for their propensity to fire not when you do something, but when *someone else* does it. This may be the hardware that goes along with this clearly important branch of learning whose purpose it is to copy, or *social learning* as I will call it.

Pointing is one of my favorite behaviors, as you presumably know from Chapter 9. Parts of pointing can be located within the social learning branch. Extended arm and index finger pointing does not occur in blind children—thanks to developmental psychologist Barbara Landau for answering this question for me. Now contrast that with motor behaviors that do occur in blind children, like smiling. In addition, the familiar pointing gesture does not occur in all cultures. Also, as noted previously, it never occurs in chimpanzees in the wild, but does in those in captivity who have had human contact. Taken together, these findings suggest that pointing is a learned behavior, despite its early appearance in infants. To be sure, the desire for shared social attention is inborn, but this particular way of achieving it is not. The familiar act of pointing that seems so human must be learned by seeing others do it—social learning.

As another aside, I have noticed that many of the reported cases of critical periods in learning seem to occur in this third branch of learning. *Critical periods* is the not-too-bad jargon for describing that sometimes proficient learning is possible only within a narrow window of time. A pre-adolescent child coming to a new country does so much better at learning the language than her mother. Dance lessons at age 15 are a lot less likely to produce a professional ballet dancer than at age 5. Both language and complex motors skills are prime examples of this third social learning branch. If critical periods in humans really do exist (see below), more work is needed to determine if this observation about their special connection to the social learning branch proves to be sound.

It is the perceptual learning branch, however, that has been the *raison d'être* of this book. It is distinct from the other branches of learning. Its evolutionary function differs in that it corrects internal errors rather than apprehends new information about the world in the case of world learning (or imitation in social learning). The conditions necessary to produce learning are different, too, because only discrepancies that implicate oneself as the blame-worthy cause of the mistake will lead to perceptual learning while other discrepancies will lead to other learning processes. The rules of learning are clearly different as demonstrated earlier by the experiment with the side-by-side comparison of perceptual learning and world learning (see Chapter 9). How learning is manifested also differs. World learning can change knowledge, belief, reflexes, emotions, memories and so on, as touched on above, but never perception, which is instead the hallmark of perceptual learning. These items (function, input, rules, output) are many of the criteria that should be different for different learning processes and perceptual learning distinguishes itself on all of them.

Perceptual Paradox

I have challenged the natural intuition that perception is normally stable. At the beginning of this book, though I prefer to think of it as a journey, perception may have seemed the safe rigid rock from which to muster a wind-blown world. Avoid loud noise, make sure the eyes are getting enough Vitamin A, but otherwise take the senses for granted. We have seen this is false. But now I will go back to the beginning and confess that the naïve instinct was not so crazy after all. Yes, perception changes with experience. It must. But we would not want perception to change with every new experience. If perception were that fickle, the same objects would be seen to be different from one instant to the next. Everything would look and sound and taste like it were transforming continuously. To hang onto a stable world, perception should only change when the underlying processes giving rise to a percept are not functioning properly. Otherwise, there would not be a need for perception to change. Moreover, such change would be harmful. Consequently, the experiences that lead to perceptual learning must imply that perceptual systems are not functioning correctly. However, this leads to what appears to be paradox.

All learning, including perceptual learning, occurs because of experience with the world. Experience with the world comes through our sense modalities. In order for perceptual learning specifically to occur, there must be evidence of an internal error. But how can information that comes through the senses possibly indicate that those very senses are incorrect? Why shouldn't that information simply be interpreted as reflecting something new, and possibly rather exciting, to be learned about the world?

This appears to be a paradox: Drifting, growing sensory systems cannot be fixed without information from the world but information from the world cannot be obtained without the sensory systems. The way out of the trap is that not all information about the world comes through our senses. Knowledge about the world also comes through our genes. Innate knowledge about the way the world works is available before any personal experience with the world. If information gathered through the sense modalities appears to differ from that knowledge then we conclude there us something wrong with the sense modalities, that is, wrong with *us*. Otherwise there would not be a way to know that our sensory systems are not working properly, to know that we have made a perceptual mistake.

We have already seen this in action. In adaptation, the perception of space changes because of the *a priori* knowledge that an object cannot be in more than one place at the same time. Provided that all the information is judged to refer to only one object, the one object-one place-one time knowledge means that the differing locations provided by the different sense modalities, vision and proprioception, would be attributed to an error in the modalities themselves. It would not be interpreted as a discovery that your cell phone can be in two places simultaneously. Whenever perceptual learning occurs, be on the lookout for the pre-existing knowledge that caused the unexpected sensory information to be interpreted as an internal error rather than as a new discovery worthy of writing home about.

Despite the fact that only very specific conditions must be met to obtain changes to perception, it would be a mistake to think of perceptual learning as a mechanism that gets turned on when needed and lies dormant otherwise. Throughout the book, we have encountered at least four things that show otherwise. We are constantly moving our bodies (Chapters 2 and 9) and time (Chapter 10) passes relentlessly. Both of these events throw off the delicate calibrations of perceptual hardware. Adaptation also maintains the body schema (Chapters 3 and 9) and maintains perceptual constancy (to be considered shortly), both needed continuously. Moreover, whether two stimuli should be perceived as the same or not (Chapters 5-7, see also Chapter 8) also continuously changes and this is at the heart of the perceptual expertise type of perceptual learning. Perceptual learning is inseparable from perception.

Hidden Error

If it ain't broke, don't fix it. Detection of what is broken has been a central part of the discussions of the perceptual adaptation type of perceptual learning all along. A modality is judged to have made an error because two modalities conflict, a conflict which is itself in conflict with certain knowledge of the world. Adaptation caused by prism-displaced vision was an excellent example. If the "ain't broke

Figure 12-3 Sabrina Peterson in Hollywood. Photograph by Garrett Richardson

don't fix it" analysis is the right one for all of perceptual learning then it must also apply to perceptual expertise, the other type of perceptual learning. Discussions of perceptual expertise have not included error detection. In perceptual expertise, we have seen that improvement in perception occurs with practice, attention, and the rest of the acquisition principles that were extracted from assorted studies. It leads to becoming better at distinguishing among faces or wines or what have you and now, as an expert, an ability to detect things never before perceived. Sounds great, but where is the error?

I suggest there has been an error all along. The hidden error involves the conclusion that there must be something there to be detected but it has failed to be detected. The failure is attributed to oneself rather than something else, like bad lighting or a trick. It's me, not you. Take a look at the picture of Sabrina in Figure 12-3 (previous page). Sabrina was my student that assisted with the mindfulness and healing studies (Chapter 11). There is another picture of her in Figure 12-4. Okay, now you have seen her. But now I will tell you that Figure 12-4 is a photo of Val, Sabrina's twin sister. Or is it? And now you have reason to try to distinguish one from the other. (Answer: It is. See the sisters side by side in Figure 12-5). The reason you can eventually tell two similar girls apart or two similar sheep apart or two similar whatevers apart is that perceptual expertise comes into play once it is realized that there are two different girls or sheep, not one.

The error detection that drives perceptual expertise then is as follows: 1) There are two stimuli, whether they be sheep, wine, chest X-rays, etc., that you perceive to be identical. 2) You have other information that indicates that they are not identical. 3) If the other information is trusted then perceiving them as the same when they are different reflects an internal error. 4) What we call perceptual expertise is the change within the modality that corrects the error and allows the difference to be perceived. There may be other things that are needed for perceptual expertise, too, but these four are

Figure 12-4 And another snapshot; see text for details.

necessary conditions.

Identifying the hidden error also means that we should dispense with the last part of the working definition of perceptual learning: Any change in perception that results from experience that removes an internal error *or otherwise improves perception.* That leaves the definition: *Perceptual learning is any change in perception that results from experience that removes an internal error.* There is no "otherwise improves perception". Perception will not improve without the detection of an error. Perception is lazy. Or perhaps cautious and efficient. The last part of the definition provided some wiggle room until all phenomena could be analyzed.

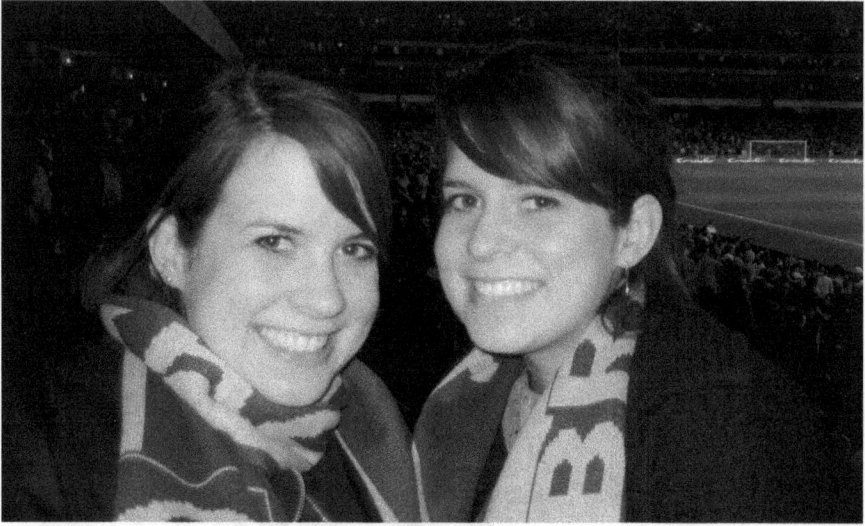

Figure 2-5 Here are twins Sabrina and Val together, but I do not know which is which... Photograph by Daniel Wood (Figure 12-4 photograph by Sabrina Peterson)

Perceptual Expertise vs. Perceptual Adaptation

We have now seen both of the sub branches of perceptual learning: perceptual expertise and perceptual adaptation. I do not mean to imply that I just picked two to discuss amidst a zillion perceptual learning processes. These are the Big Two. Table 12-1 shows how they compare on key features. The two processes together do a pretty good job at covering the universe of possible changes to perception. Some of the ways in which they have complementary properties is satisfying. For instance, perceptual adaptation is about changing whole perceptual dimensions, like a spatial dimension from left to right (azimuth angle), while perceptual expertise is about changing specific stimuli within those dimension, like a stimulus with a particular orientation in space. Perceptual expertise requires attention to what has to be learned whereas perceptual adaptation does not. Perceptual adaptation requires feedback but perceptual expertise does not. In perceptual adaptation, one object or event erroneously gets assigned two different values initially whereas in perceptual expertise, it is two objects or events that erroneously get assigned one value initially. In other ways, they are alike. For both kinds of learn

Table 12-1 Comparison of Perceptual Learning Types

	Perceptual Expertise	Perceptual Adaptation
What results from **Learning**?	Detect the difference between perceptual stimuli	Recalibrate a perceptual dimension
Does what is learned **Generalize**?	No. Specific to trained stimulus; i.e., discriminates (may "generalize" negatively)	Yes. Generalizes to untrained stimuli along the same dimension
Is **Attention** required for learning?	Yes. Requires attention to the specific feature to be discriminated	No. Does not require attention to the conflict but what is attended can influence what changes
What is the nature of the **Error/Discrepancy**?	Two different objects get one percept when there should be two percepts	One object gets two different percepts when there should be one percept
Is the *Object Identity* decision required?	Yes. To get learning, must conclude first that there are *two* different objects	Yes. To get learning, must conclude first that there is only *one* object
How many **Acquisition** trials are needed to get learning?	Hundreds and thousands but can be reduced dramatically, even to one trial, if you know what you are doing	Hours and days of running around the world, but only five minutes if you know what you are doing
Is **Feedback** needed to learn?	No. But it may help in ways that are not yet clear.	Yes. Must be checked against something else to get learning, though not conscious feedback.
What is the **A Priori Knowledge** that causes an internal error to be detected?	?	Varies; e.g., One object can only be in one place at one time; e.g., An object should

		not change color when its orientation with respect to the perceiver changes
What is the **Mapping** between the physical world and the psychological one?	Go from a many-to-one mapping before learning to a one-to-one mapping after (but other characterizations possible)	Go from a one to one mapping to a different one to one mapping that is often linearly related to the first
Is there a **Statistical Error** that is relevant?	Reduction in variable error $VE = (\Sigma (Xi - M)^2)/n$ where Xi = perceived value on Trial i M = Average across i trials n = number of trials	Reduction in constant error (usually) $CE = (\Sigma (Xi - D))/n$ Xi = perceived value on Trial i D = actual value n = number of trials
Why does the mechanism exist?	Because whether two stimuli are the same or different changes depending on the context	Because perceptual systems drift from accuracy with passage of time, because of observer movement, and maybe growth
What are some **Examples**?	Distinguish between two oriented lines, two graham crackers, two faces, two perfumes	Shift perception of time, space, color, size
What are some **Uses**?	Acquire professional skills, pleasure of perceiving new things, detect things you need in your environment	Keep body schema accurate, maintain perceptual constancy, physical healing
Is there a **Critical Period** of life within which more learning is possible?	Unknown, but lifelong learning is possible	Unknown, but lifelong learning is possible

Table 12-1 continued Comparison of perceptual learning types

ing, the amount of practice needed can be reduced dramatically by explicitly concentrating the conditions that trigger the changes. They both require the object identity decision in which two samples are attributed to either one object or two. And, of course, they both lead to changes in actual perception.

Figure 12-2 also shows some branches of each perceptual learning process. See the three chapters devoted to each process for the details of the examples, including some deep points about perceptual learning. Does every last change to perception fall into one of these two types? Some effects that seem to require their own process turn out to be just a part of one of these. For instance, perceptual categorization, in which two different items are lumped together, is really just perceptual expertise in reverse. In the section "Loose Ends" below I show how other changes to perception are well accommodated by these two processes. I will go out on a limb of our tree of learning and say "yes", I believe these are the only two basic processes. I could be wrong, in which case I am not done. Check the bookshelves for a sequel.

Time Course of Perceptual Learning

I have not said much about how long changes to perception last. That is because every time I have looked, they have lasted longer than has been reported. Perceptual learning does not "decay"[2]. If a change is adaptive then it is kept and if something else is more adaptive, it is replaced. Some situations will demand very rapid replacement while other changes could remain well-suited to an environment for a lifetime. There have been attempts to classify only changes that last "relatively long" or are "semi-permanent" as true learning with more ephemeral affects attributed to a mysterious something else. I think this line of reasoning is incorrect. The analysis of learning as those products of experience that make one better (Chapter 1) does a better job of capturing the intuition of what is and is not learning. I suspect that the longevity of change was a roundabout way of trying to get at the real underlying issue of making us better without ever quite making it, with effects that didn't last very long either observed or surmised to be less likely to do much for our improvement. There is no additional requirement on top of improvement and no set time to how long a change must last to be learning.

While changes will not diminish if not replaced with something more adaptive, items learned both before and after may interfere with what is retained, as it does in all learning processes. Often there is an overlooked advantage to such interference and a replacement is more adaptive than it at first seems. A related matter is that sometimes the experience producing the learning has not been rich enough to specify the context in which the change is appropriate. When this happens, a change may appear to drift back to the original state and is really due to two conflicting changes, one new, one old, either one of which may be appropriate for the current situation. In addition, cell death can result in loss of what is learned through aging or other damage but this is true for all biological systems and not just perceptual learning. As noted before, experience can make us worse as well as better.

Similarly, when I have investigated claims of something perceptual taking an unusually long time to acquire compared to other kinds of learning, I have found it can be sped up considerably. This can be done by ensuring the exact requirements for getting perception to change are met. Just like on the back end where there is no set requirement for how long a change must last to be considered true learning, I argue there is also no set requirement on the front end for how long something has to take to acquire to be considered true learning: one trial or a thousand, depending, in part, on the care with which the training trials are selected.

I have not investigated claims of critical periods in perceptual learning but I would not be surprised to find that here, too, the timing can be manipulated if you know what you are doing. An adult might require a greater amount of isolation of the feature to be discriminated than a child in order to get the same degree of perceptual expertise, for example, because a greater prior experience with other features will make the new feature harder to attend to. This could masquerade as a critical period in which learning only seemed possible in early childhood.

Loose Ends—McCollough Effect and Other Perceptual Phenomena

It is worth having a look at other changes in perception to assess how well they are captured by the analyses both of what perceptual learning is about and its division into two types. At the top of this list is the *McCollough Effect*, a popular illusion in perception. In the McCollough Effect, viewing red (magenta works better) vertical lines alternating with green horizontal lines for a few minutes leads to illusory color perception in which vertical lines now look faintly green and white horizontal lines look pink. Figure 12-6 shows what typical stimuli look like. To try the illusion for yourself though you

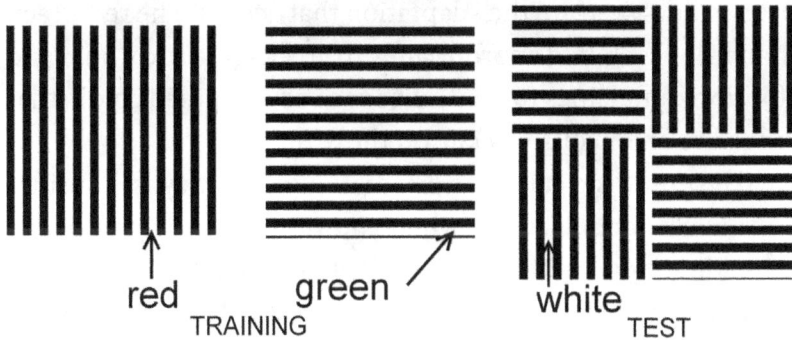

red   green   white
TRAINING   TEST

Figure 12-6 The McCollough Effect stimuli, but shown in black and white. Subjects see red vertical lines alternating with green horizontal lines. This leads to an illusion in which white horizontal lines look light green and vertical lines, pink. In the training figures above, the vertical grating would be colored red and black not white and black, and the horizontal grating green and black. (The test grating would be white and black as shown.)

will need to view it in color which you can do on the web. This is a very long lasting illusion—days, sometimes longer (possibly forever if you never see vertical or horizontal lines of a different color; see section preceding on the time course of perceptual learning.) The bias I mentioned earlier in which researchers ascribed long-lasting changes to learning caused quite a number of people from different areas to sit up and take notice of the McCollough Effect. There is clearly a change in perception: The exact same stimulus (white vertical lines) appears one way before training (white) and a different way (greenish) after experience. It meets our t1-experience-t2 framework. If this is perceptual learning, what is it about the displays that indicate that our perceptual systems are malfunctioning?

I have argued for many years that the vertical and the horizontal displays of the McCollough Effect indicate that a single object—a grating of lines—is changing color when the orientation of the object changes on the retina. Yet an object should not change its color when the retinal orientation changes. To allow this to happen would undermine *orientation constancy.* Consequently, the change of color with a change of orientation is attributed to a perceptual error and the illusory colors seen are the adaptation that occurs. The red-green color dimension is shifted appropriately to green for vertical lines and to red for horizontal lines to neutralize the red and green colors that were judged to erroneously occur on those lines.

To make sure it is clear what orientation constancy is, consider that in ordinary circumstances when you tilt your head, the object you are looking at does not appear to tilt. This is notable because the orientation of the object on the retina has tilted. Tilt your head from vertically upright to horizontally on its side and the retinal orientation of a television you are watching changes from vertical to horizontal yet the television continues to look vertical, not horizontal. The actual orientation of the object stays constant; you achieve orientation constancy and perceive it accurately as unmoving. (If you have never noticed this before, try tilting your head now and pay attention to how the objects do not appear to change orientation. If you pay close enough attention, you can also become aware that the retinal image is different. If the orientation of the object ever did appear to change then you *would* have noticed, so it is a very good sign that you never gave it much thought. Now try pushing gently on your eye with your finger and the world does appear to move because in this situation, you do not achieve constancy, in this case motion constancy.) When the orientation of an object changes on your retina but the object didn't budge, you don't want to see it as changing because if you did you would be wrong. Now, just as you don't want to see the object as tilting just because the retinal orientation changes, you also don't want to see its color changing just because the retinal orientation changes. If you did, you would be wrong. The adaptation that we call the McCollough Effect fixes that error if you do. I like to think of this as *secondary constancy,* while the more direct traditional constancies, like orientation constancy in spite of retinal orientation changes, are primary.

In space adaptation, there were real world circumstances outside the lab that would cause an object to *appear* to be in two places simultaneously when assessed by two different modalities, even though it is impossible that it really is. Thus, having a mechanism ready to correct for this error was sensible. In the McCollough Effect, there must also be a reason to be ready to correct for something that doesn't happen. There must be some situation in which it could appear as if the color were changing when the orientation of the object on the retina changed, just like it could appear there was a violation of the one-object one place one time constraint in space adaptation. It turns out there is: Unless the mechanics of the eye are perfect, color fringes can appear at edges. The eye acts like a prism and disperses the different wavelengths of light. Richard Held gets the credit for pointing out that the optic axis of the eye has to be aligned with the fixation axis of the eye or else you get these color fringes, although he didn't make a big deal out of it and only discussed it once.

Do you remember that only long lasting effects were falsely viewed as "real learning" and that, in addition, "real learning" was falsely restricted to world learning? As a result, it is a common error (by people this time, not the perceptual system!) to attribute the McCollough Effect to world learning rather than perceptual learning. One of the most popular world learning accounts maintain that observers learn that vertical in the world is associated with red in the world. The mechanism suggested for this is either Pavlovian conditioning or, less often, through one that tallies frequencies of co-occurrences of red-vertical objects in the world[3]. But as we have seen, the rules that govern changes to actual perception are different than those that are operating to acquire new knowledge about the world. Therefore, these theories are wrong. The McCollough Effect and its many relatives (like color contingent on the direction of motion instead of the orientation or motion contingent on the width of lines) are about correcting internal errors within perceptual dimensions in order to maintain perceptual constancy, They do not reflect learning new things about the world. The McCollough Effect is the perceptual adaptation type of perceptual learning with shifts in a color dimension rather than shifts in a space or time dimension. I have written at length elsewhere about the details[2].

I visited Celeste McCollough once at the Williams Air Force Base in Arizona, nearly three decades after she discovered the effect that bears her name to this day. She had become Celeste Howard fairly soon after her discovery and I guess titles don't transfer since it did not become the Howard effect. It also meant a lot of people couldn't find her work until much later when she started using both names. She scoffed at my theory by the way. She did not find an interpretation based on perceptual errors to be correct. She stuck with her original view that the effect pointed to bipolar neurons that coded for both orientation and color, such as a neuron that responded to vertical gratings only when they were colored green. How they are coded neurally seemed to me beside the point (and still does); we still needed to know the function of the McCollough effect, whether there exist similar effects, and other related psychological questions. She was delightful though (complete with a chopstick in her hair) and helped measure the colors on the computer monitor that a student, Karen Reinke, and I were using for experiments. A gracious host.

In the McCollough Effect, the error is detected entirely within vision. Contrast that with space and time adaptation, where the discrepant information came from two different modalities with vision indicating it was one way, proprioception another. There are other examples of perceptual adaptation besides the McCollough Effect where the discrepant values can come from within the same modality. Perceiving how far away an object is can be achieved with about a dozen different *depth cues*, all within vision. These include *binocular disparity* (which uses the fact that objects at different distances cast images on each eye that are a different distance apart) and *texture gradients* (things far away appear cast smaller, more densely, packed images that are higher up than things close by). Each cue tends to have a specialty, but there is plenty of overlap in what they do, just as both vision and proprioception overlap in the ability to determine the angle at which an object is located. By the way, *perception is as redundant as it is lazy.* A clever intervention to make depth cues systematically conflict with one another causes a shift in one of the depth cues, just as cross-modal conflicts cause a shift in one of the modalities. Hans Wallach showed that.

The conflict I would like to see created that is entirely within a single modality is to pit the two different ways of judging the location of an object within vision against one another. I had mentioned something so quickly in a previous chapter that it probably went unnoticed. When I explained how we can determine through our eyes what the angle of an object is—Is it straight ahead? Or 30° to my left? All the way to the left?—I described how the retinal position of the object is combined with position of the eyes in the head (and head on shoulders if need be) to calculate the object's true location in space. But I also mentioned that there was another way to do so if the room lights were on. The other way involves comparing the location of an object to the larger environment one sees. If a small dot is centered in a large frame, then *that position* is straight ahead. If it is at the left edge, then the dot is localized all the way to the left. Discovery of the former calculation (*unconscious inference*) system, or egocentric perception, was by Herman von Helmholtz (who also gets credited with being the first to play with prisms, see Chapter 9) and the latter relative comparison system, or exocentric perception, is due to James Gibson (Chapter 6). I would love to pit these two systems against one another by making them conflict about the location of an object and see who wins, Helmholtz or Gibson[4]. The metaphorical conflict this would create between two larger-than-life theorists that shaped the field of perception is too hard to resist.

Sometimes, there even appears to be adaptation-like changes when only one system of any kind is apparent and not even two systems within the same modality. These can be harder to identify where the conflict is lurking, but not impossible. There is a many decades old Gibson-like phenomenon in perception that was called *adaptation level,* another unfortunate, confusing label. Take any single perceived dimension, like weight. It turns out you can move around how heavy an item actually feels to you just by changing the weight of other items in the environment. A 3 pound pumpkin lifted following experience with multiple items weighing between ½ pound and 2 pounds will feel heavier than the pumpkin following experience with 4 to 10 lbs items instead. The experience causes a long-lasting shift in the entire perceived weight dimension—this sure looks like adaptation. Likewise, 85° feels warm in early June in Brooklyn, New York but feels a lot cooler in Tucson, Arizona (even with the same humidity!) because of all the preceding scorching days. The straight-ahead shift mentioned in Chapter 9, the seeming imposter to genuine space adaptation, provides another example. If items you see in your world are no longer evenly distributed around you but are instead consistently more on the right side than your left, then what looks to you to be straight ahead shifts to a position further to the right. These shifts following lopsided experience pretty much happens to any perceptual dimension you can think of. But it is still just experience with *one* perceptual dimension. Without anything to compare it to, how is the dimension recalibrating?

There are a few different possibilities. One is that the dimension is compared to another perceptual dimension *from a different time* rather than to another modality present at the same time or another system within the same modality (at the same time). The weight dimension from now is compared to the weight dimension from yesterday—those are two different dimensions—and if it differs, there is a conflict. One or the other must shift then, just as with any other conflict. If there is a conflict, can the time dimension from yesterday shift rather than the one from right now? I am not sure. Like with time adaptation discussed in Chapter 10, we are not entirely comfortable thinking of different temporal "places" getting compared and have an easier time with different spatial places. Another possibility is that the dimension is compared to itself at the same time. Some dimensions are symmetrical, like angular position in space—the left side is similar to the right. Can the right and left sides be compared to look for a mismatch? I think whatever comparison is available, perceptual systems make use of it. Otherwise, we run the risk of making errors and not being able to fix them.

Incidentally, to be thorough, let me clarify one thing. When discussing Harris's "beware the straight ahead shift" as an imposter to adaptation (Chapter 9), a more precise description is that it is an imposter to *prism* adaptation. The straight-ahead shift may also be adaptation, but not one that needs, or is created by, a conflict between sensory modalities.

There are also changes to perception from experience known as simple aftereffects and some of them as afterimages. Have you ever stared at a waterfall for a minute and then found that stationary items look like they are moving up? Or seen blue blobs after staring at a yellow light source? If these changes are merely fatigue in sensory receptors and they do not make us better then they are not perceptual learning at all and need not be considered further. If on the other hand, fatigue itself turns out to make us better, perhaps by preventing injury, or if simple aftereffects reflect a change to fix some hidden perceptual error, then simple aftereffects would be perceptual learning and we would need to locate them on the tree. Likely adaptation will be the helpful mechanism if it turns out that these effects are indeed learning.

To pause briefly, the point of much of this discussion is to illustrate that the Big Two perceptual learning processes appear to have great potential for explaining the vast number of ways in which perception is pushed around by experience.

To resume being thorough in this loose ends section, let's also be sure that the findings from the first part of this book can be accommodated within our understanding of these perceptual learning mechanisms. The second part of the book dealt explicitly with the perceptual expertise branch of perceptual learning (Chapters 5–7) and third part with perceptual adaptation (Chapters 8–11). But the first part covered topics with the intent of introducing both the idea and plausibility that perception changes rather than identifying the type of perceptual learning responsible. We have already located damage to perception (Chapter 1) on the tree of experience. Issues of growth from Chapter 2 were considered in later discussions of adaptation. In addition, there is a simultaneous component of expertise in growth, as also noted Chapter 2, as when the precision of sound localization increases with growth. Changes to body schema with tool use, the subject of Chapter 3, also meshes well (pun anticipated?) with adaptation: That tennis racquet you don't let go of for an hour provides conflicting information to sense modalities. Vision and the motor system signal it is part of your arm—you see it physically connected to your arm and you use it the way you use your arm —yet proprioception has no representation of it. That is, you don't feel the racquet itself, just where you grip it. The conflict may be resolved by extending where the hand is represented to have its end to include the racquet, perhaps a simple magnification change in the length of the felt arm, given the preference of adaptation systems for linear changes. It is a little tricky because you can never actually have proprioceptive receptors on the tennis racquet itself. Yet you can still have a "felt position sense of tennis racquet" if the tip is touched and vibrations carry that to your real hand where there are receptors. Much like damage to a limb might cause proprioceptors, and hence felt position sense signals from the limb, to be lost, we need not necessarily deny ownership of that limb (though sometimes this can happen). More work is needed, but you get the idea: Adaptation generally seems like a promising analysis for incorporating tools into the body schema.

In the final introductory chapter, Chapter 4, subjects wielded tools to determine their length and width dimensions without getting to use vision. Perception of those dimensions changed with experience. As noted in that chapter when previewing what was to come later in the book, part of that learning reflects adaptation (the reduction in constant error) and part reflects perceptual expertise (the reduction in variable error). Adaptation causes the shifts in the perception of length and width if the subject had previously over or underestimated while expertise causes a greater precision with which length and width is perceived, if the subject's judgments were initially very variable.

I believe that brings us up to speed on how to incorporate the introductory chapters.

Another loose end: My use of the framework t1-experience-t2 diminished as the discussion advanced. That is not because I decided the framework was not useful after all. Rather, as the examples become more obviously perceptual learning, there was less of a need to verify that with the framework. I continue to believe the framework can play a useful role in perceptual learning, especially for examples where it is not clear whether perceptual learning has taken place. If I also decreased the frequency with which take-home messages were italicized, it is because they became too dense in later chapters to highlight them all.

For the final loose end, see Figure 12-7. That's Tony.

Figure 12-7 Tony with and without hair. Here he has the same expression, but in class he looked like two completely different people. We do not want to perceive the zillions of different possible Tony images and therefore are often better off not seeing differences at all. See Chapters 5 & 6. Cartoon created by students in response to Tony issues. It was a web class and she didn't know what I looked like...

Photos courtesy of Tony Golembiewski. Original cartoon by Katelyn Becker and Keisha Everett.

Terminology (mostly)

Know. Believe. Think. Judge. Interpret. I have used all of these terms and more when describing what both perceivers and perceptual systems do. I hope I did not leave the impression that conscious reasoning intervenes between the stimulus on the sensory receptor and the eventual percept because that was not my intent. The problem is that there is no vocabulary that adequately captures the perceptual chain of events without using words that ordinarily imply conscious deliberation. The alternative is to use neutral phrases like "the output of the perceptual system" but that gets cumbersome, snooze-worthy, and even meaningless, very quickly. Hopefully, on rereading, it will be clear that conscious awareness is not the intended meaning behind the cognitively-laden terminology.

The issue of conscious awareness itself is, of course, very interesting. I have sprinkled throughout comments on consciousness including: It co-occurs frequently with "the output of a perceptual system", but need not, that internal perceptual wars (such as those between sense modalities) occur outside of conscious awareness, and that thoughts typically do not influence perception (but note the object-identity decision deserves special consideration). Mostly, though, I leave analysis of conscious awareness to others. Nailing down consciousness and its ally "attention" is a bear, but it is for all of psychology. Its importance for healing (Chapter 11) may give it a new twist.

The word "perception" does not just mean vision although some people use it that way, which could be confusing. I have used *perception* to refer more generally to the output of any sense modality.

Also potentially confusing is that some people only use "learning" to refer to what we have called "world learning", and even worse, the rat in maze people co-opted the term "learning" to mean only Pavlovian or instrumental conditioning. This was raised earlier but is worth reiterating in a discussion about terminology. Both views of learning are artificially restrictive and I have used learning throughout in a much broader sense.

Likewise, I have been repeating that the perceptual expertise people artificially restricted the term "perceptual learning" to one branch of perceptual learning much like the conditioning associationists did for the term "learning". I have used "perceptual expertise" as a fitting label to refer to this type of perceptual change. Unfortunately, the old-fashioned use of the term "perceptual learning" in the literature still leads to a great deal of confusion. A reminder here then that "perceptual learning" is used throughout as the over-arching umbrella term. With the qualifications and tweaking throughout, we are left with:

Perceptual learning can be said to occur when, and only when, there is any change in actual perception—literally seeing, hearing, touching, smelling, tasting, and immunating—that is caused by experience and serves to remove internal errors through the detection of errors and correction of the perceptual system inferred to be the cause of each error.

FUTURE

The human genome project was completed in the year 2003. Humans possess some 3 billion pairs of nucleotide bases of guanine (G), thymine (T), adenine (A) and cytosine (C). Strings of these comprising about 20,000 genes that vary in length from only 251 bases all the way to 2.4 million bases are the design plans for all that you are born with. The genes are located on 23 pairs of socks, I mean chromosomes, but they are matched pairs like socks, in the nucleus of each cell. Each chromosome contains between 231 and 2968 genes (DNA in the mitochondria rather than the nucleus adds an additional 37 genes with 16,569 positions). Moreover, the cost of sequencing a person's entire 6 billion "letters"—those 3 billion pairs—has fallen to $1000 and may continue to fall. Even more practically, any individual can choose right now to get one million positions along their genome tested for the cost of only a month or two of cable TV. These million positions tested contain sites of high variability between people and therefore provide information on how you differ from others—disease risk, ancestry, and all manners of traits, physical and psychological.

This is an amazing milestone in humanity's history. I hope it will forever change the future of science, including psychology and perception. There are two notable ways in which the advancement of genetics should bear on perceptual learning. My favorite is about the detection of errors. We have seen that perceptual learning is all about detecting and correcting internal errors. Errors are also important in DNA. When DNA is copied to make new cells, mistakes can occur. A string of CCCC for example, may appear as CCCCC in the copy (think of this as counting swim laps; was that just my 4th or my 5th?), or a letter might get deleted rather than inserted (CCC), or one substituted for another, such as a "T" for a "C" (CTCC). Given the amount of copying that has to occur and the dramatic effect a change in life's blueprint can have, detecting and repairing DNA becomes critical to minimize damage. When there are two copies of a section of DNA, one you got from your mother and one from your father, they can be compared to one another and differences detected. Does this sound at all like comparing one modality to another in perceptual adaptation to assess agreement on a perceptual feature? Note there also may be an analogous problem to figuring out which modality is right of figuring out which copy is right. There is only one copy though for the Y chromosome, which gets passed only from fathers to sons, and for DNA in the mitochondria, which both boys and girls get only from their mother. With only one copy, error detection is more challenging and indeed, less effective. Within the Y chromosome, there is a certain amount of symmetry and one end of the Y is compared to the other end. Error detection within a single dimension also occurs in adaptation when there is no other system to compare it to.

There may be only so many fundamental ways that errors can be detected. This may have led to the same types of processes having evolved to manage errors both for perception and for the stuff of life itself. Genetics and perceptual learning, then, may share a special relationship.

The second major connection to perceptual learning is that genetics has always been learning's other half. It just has traditionally gone by a different name within psychology: Innate vs. Learned, Nature vs. Nurture, or Nativism vs. Empiricism. While the effects of learning on our psychological make-up could be readily studied in the lab, the innate contributions to our constitution could only be inferred through indirect means. This often meant time consuming adoption studies to try to tease apart inborn from acquired characteristics. Results of these studies were always fraught with ambiguity of interpretation. If the adoption did not occur soon enough, what looked like an innate similarity to the biological parents could really be an early learned one instead. If adoption placement was not random, what looked like a learned component could really be innate. Direct testing of genes within the individual himself or herself bypasses a lot of the mess.

For instance, there is a site on chromosome 7 in the gene known as TAS2R38 that influences the perception of bitter taste. People who have two copies of C at position rs713598 usually cannot taste the chemical propylthiouracil, which is similar to substances found in broccoli, cabbage, coffee, and dark beer. Those who have either CG or GG instead find that these substances taste unpleasantly bitter. About 20% of Americans from Northern and Western Europe have CC (my own is CG, found in 50%) and that incidence more than doubles in Han Chinese of Bejing. The genetic information can inform how much perceptual learning work will be needed if someone is to become an expert at distinguishing similar bitter tastes from one another—like a beer expert. Someone whose genes give them a bitter taste advantage would need a less intense perceptual expertise training program than someone whose genes do not. The genes provide the foundation on which to build upon with experience. Epigenetics, in which the environment in turn changes which genes are turned on and off, may illuminate an even richer way in which inborn and learned knowledge interact in perceptual learning.

Both DNA and experience give us our individual differences as well as our similarities to one another. We are currently looking for the genes that may explain the spatial problems that only some people have. Those variants were discussed in Chapter 9 and include frequently getting lost or confusing left and right. When differences have an extremely negative effect on functioning, they are often considered to be deficits rather than just variants, although the line between normal and abnormal is not always clear. If autism is perceptual expertise taken to the max, such that two different stimuli are never categorized but instead always perceived as different, then searching for a genetic perceptual basis for autism would be important. It would be of benefit both for those who have the disorder and for furthering the understanding of the perceptual expertise mechanism. The field of perception has favored general laws over individual differences, with a few exceptions for deficits like color blindness. Hopefully, genetics can lead the way to a study of individual differences in perception on par with the general laws. See also Chapter 13, "April. Test your DNA".

Stretches of DNA in which everyone is the same is also important to genetics. Non-varying genes are often referred to as highly *conserved*, meaning that they are so important to survival that any variants would not remain in the gene pool. In fact, 99.9% of all nucleotide bases are identical in everyone. In perceptual learning, we discussed how *a priori* knowledge of the world is indispensible for getting perception to change. We are born knowing that one object cannot be in more than one place at the same time. In psychology, examples like these are sometimes known as constraints. As we have seen, without some *a priori* knowledge or constraints, there would be no way to use the world to figure out when it is our own internal assessment of the world that is wrong. These constraints are probably then highly conserved in the genes, to use the terminology of genetics, because we would not last long if there were widespread inaccuracies in our senses. Perception folk have not really known what to do with "constraints" on perception and learning any more than they did with "plasticity" in perception. Instead of viewing them as constraints, or limitations, thinking about them as highly conserved properties and searching for their genetic basis may be a more fruitful approach. Re-thinking them in this way also contributes to a much needed link from perceptual learning to biology.

I am also excited for a future with the immune system as one of our long-lost senses. As developed in Chapter 11, interaction of the immune system with the other modalities, like vision, leads to a surprising outcome. What appeared to be mind over matter processes more akin to magic or the supernatural than to science can finally be demystified and understood as perceptual adaptation. Meditation and guided imagery for healing are just like adapting to those goofy prisms and just reflect the ordinary perceptual error surveillance machinery doing its thing. There is much here to explore. One example: If the immune system shows perceptual adaptation, one of our perceptual learning mechanisms, is it also subject to the other one, perceptual expertise? This would imply some kind of increased precision in the detectability and discriminability of potentially foreign invaders as a result of suitable experience. I do not know what effect that would have on physical well-being, but it is definitely worth finding out.

Further connecting perceptual learning to biology and medicine can come from separate reports that the McCollough Effect is influenced by the amount of sleep and by phase of the menstrual cycle[5]. People who have slept fewer hours the night before show less of the illusory colors on oriented lines than people who got a good night's sleep. Sleep loss is known to affect the immune system which may provide the eventual connection to perceptual learning. Alternatively, it could implicate a role for hormones, also modulated by sleep, in perceptual learning. This would fit with the changing strength of the McCollough effect with changing hormones from the other study, in which more saturated illusory colors were reported seen after ovulation when progesterone levels increase and central arousal decreases. Central nervous system arousal may also explain all the influences.

Why do these biological factors influence specifically the McCollough Effect? It may be that no one has really looked at the other types of perceptual learning for influences of hormones or immune mediators. Note I have not found any relation between number of hours of sleep and amount of adaptation to prisms, but that result was just an add-on to another study I was doing and I have not attempted to rigorously find out the answer one way or the other. Or maybe the McCollough Effect is perceptual learning's canary in a coal mine, with great sensitivity for minute changes in internal milieu. If so, that would make the McEffect, as I like to nickname it when chatting with students, an ideal paradigm to further pursue biological influences. Incidentally, I do find that women show a larger McEffect than men, so much so that I switched to using only women for those studies.

Following the lead of psychoneuroimmunology (See Chapter 11), further investigations of neurotransmitters and the nervous system in perceptual learning, as well as hormones and the immune system, would be a good idea. Note that I did not say neurons or brain regions. It is fashionable to do studies using the technology of *fMRI*, functional magnetic resonance imaging. Even major government granting agencies, like the National Science Foundation, have boarded the bandwagon and encourage such work with a much higher rate of funding. In an fMRI procedure, images of the brain are taken while subjects engage in a specific chosen task—maybe looking at a picture, or hearing a sentence, or trying to recall a list of words— to see what parts of the brain are active during the task. This reminds me of what I heard about what shoe stores were once like. They had X-ray machines so that customers could see through their feet to the bones. It was cool. You got to verify something that you know is there but that you don't ordinarily get to see. But when all is said and done, you still need shoes. Hundreds of hours of fMRI work often produce little more than a simple list that would fit on a postcard of what part of the brain does what. We already know that absolutely everything we do has to be accompanied by neurons firing in the brain some- where, so it should come as no surprise that some region of the brain or other will be active during any task you provide. Yet people none- theless still light up when they get to see those brain regions light up—it's just like customers seeing for themselves the bones in their feet. You still need shoes.

There are exceptions. I have tried to mention notable findings on neurons that go beyond listing a brain region, although none of these were discovered with fMRI: neurons that respond equally to visual stimulation on both a limb and a tool, the stealing of neurons from other brain regions that occurs in perceptual expertise, mirror neurons that respond when someone else acts. I do not think future advancement lies with mapping out brain regions. Remember, 90% of faculty in psychology departments were once upon a time running rats in Skinner boxes. Now, conditioning is sometimes not taught at all in psychology departments. Shiny new toys come and go. Years ago when I wrote about how the Not-the-Liver Fallacy was producing artificial results[6], I predicted that "implicit memory", which at the time was causing cognitive psychology to be all abuzz, would vanish. I also said I hoped it would happen in my lifetime. It happened pretty fast and I have not even heard a single colloquium on the topic in the last five years. I am not sure I will get as lucky with the brain craze; I'm older. But the prediction is the same.

Did you know that fewer students enroll in perception classes than any other psychology class? I think it is because there is too much of a disconnect between the abstract theoretical concepts in perception and a person's real life. The fact that perception takes place so fast and out of awareness doesn't help any either. But I think that trend can be not only changed, but reversed. Since our experience of the world has to come through the senses, this gives perception the strongest connection to real life, not the weakest. I would like to think that demystifying the supernatural with perception as I have mentioned throughout this work—out of body experiences and astral travel, spirits stealing favorite objects, healing the body with the mind, déjà vu—may entice people to have a second "look" ("see", we can't even talk about anything without referring to perception!) at the field of perception. When I was in the sixth grade, I read the book A *Pocket Guide to the Supernatural*[7]. It would be great to re-write that from a perception perspective. I would also like to try some new crazy things, like trying to give ordinary people synesthesia so that they can hear tastes or see sounds, building rooms with duplicates so that it looks like every object is in two places at the same time, and exploring mentally travelling in time. In any event, applying perceptual learning to real-life concerns like perceiving faces of other races, showing up late for appointments, understanding the spatially challenged among us, and using perception to improve health, are baby steps in the right direction.

This is it. (almost)

Experience

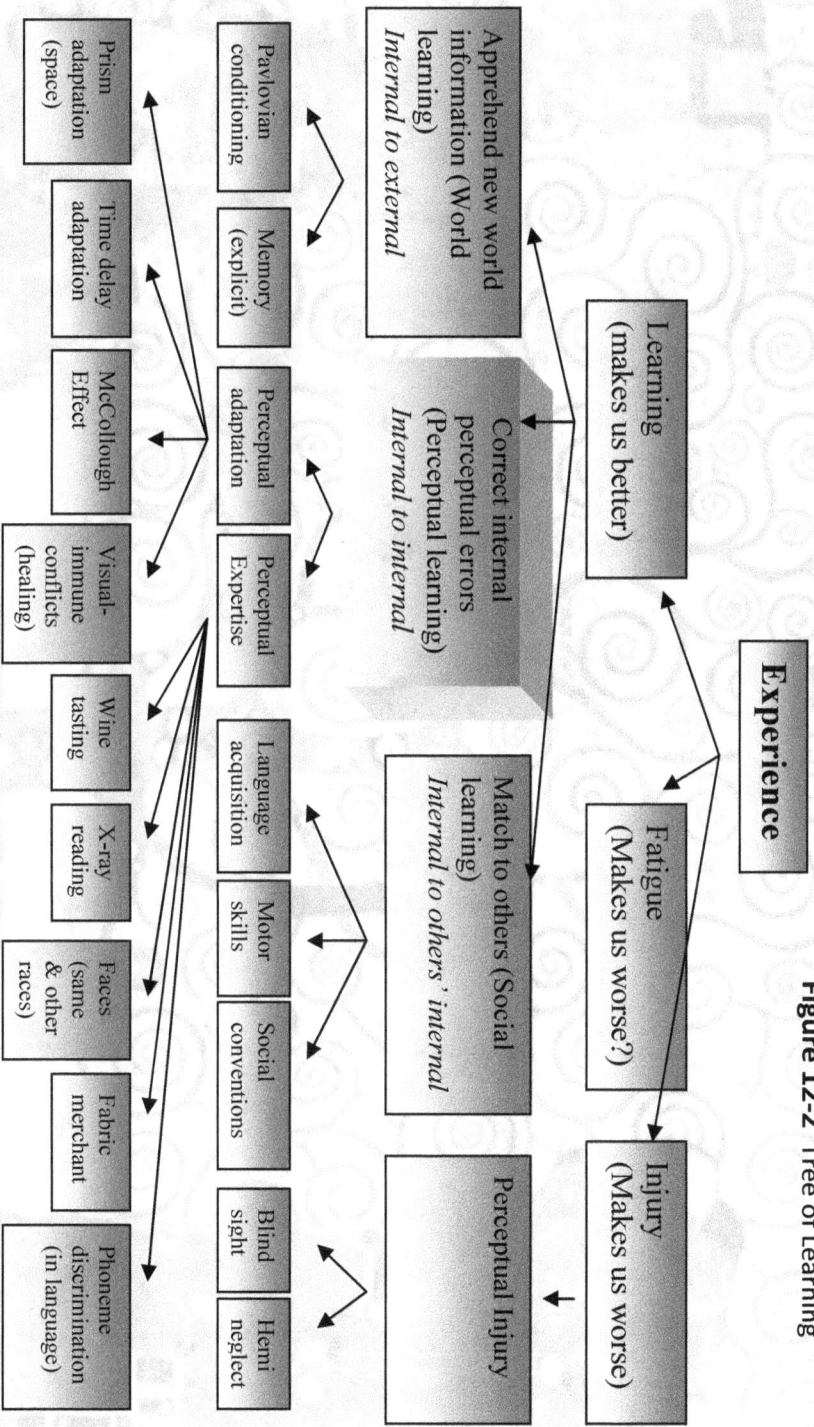

Figure 12-2 Tree of Learning

Learning (makes us better)

Fatigue (Makes us worse?)

Injury (Makes us worse)

Apprehend new world information (World learning) Internal to external

Correct internal perceptual errors (Perceptual learning) Internal to internal

Match to others (Social learning) Internal to others' internal

Perceptual Injury

Pavlovian conditioning

Memory (explicit)

Perceptual adaptation

Perceptual Expertise

Language acquisition

Motor skills

Social conventions

Blind sight

Hemi neglect

Prism adaptation (space)

Time delay adaptation

McCollough Effect

Visual-immune conflicts (healing)

Wine tasting

X-ray reading

Faces (same & other races)

Fabric merchant

Phoneme discrimination (in language)

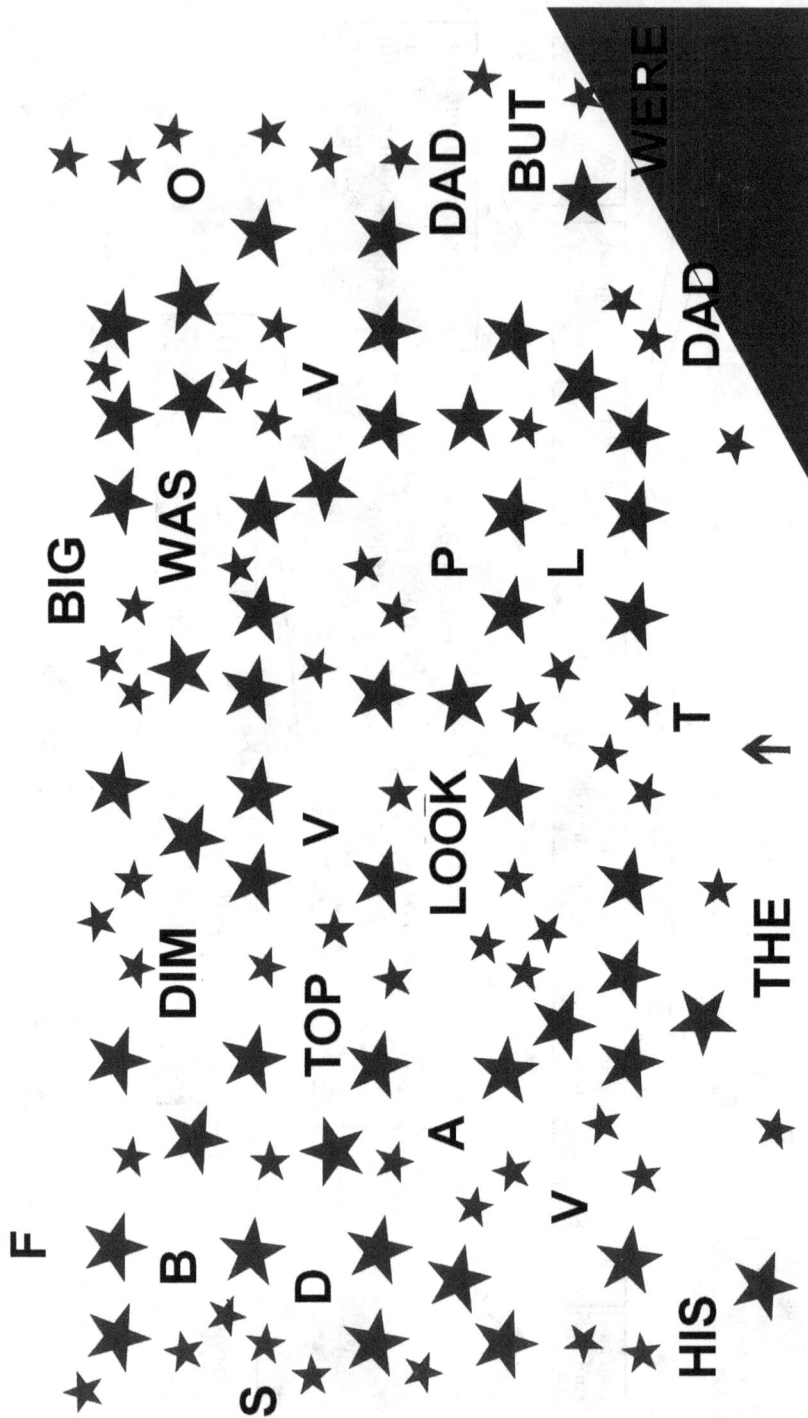

Figure 13-1 Try your hand and eye at this variant of the Star Cancellation Test. Center yourself where the arrow is (bottom, middle). Now cross out all the small stars. Adapted from the Star Cancellation subtest of the Behavioural Inattention Test. Copyright © 1987 & 2012 by Barbara A Wilson, Janet Cockburn and Peter W Halligan. All rights reserved. Adapted with permission of Pearson Education Limited. Part of the content of the image from the actual assessment only have been shown above in order to protect the security of the test.

How do YOU Measure Up?

Now you, too, can use learning to become a perceptual super-hero. Why bother? You will get more enjoyment out of life because all life comes through the senses. Experiences will be more intense. Food will taste better. You can identify perceptual blind spots. Improve your well being with less pain and more relaxation. Learn to be on time for appointments, become a birdwatcher, recognize faces like never before. Plus, perception is fun. Here are a few to get you started.

January Don't Ignore the Left Side Of Space!

Do you have a left or right side bias that has some of the same properties as the perceptual disorder hemineglect (Chapter 1)? You may and not even know. Try the star cancellation test[1] in Figure 13-1. Set the book down in one place and then cross out all the small stars in the illustration. Even better is to make a video of yourself doing the task to watch afterwards. Come back here when you are done. One more thing to try: The next 5 times you park your car in a shopping center, record whether you turned left or right when there was a choice. Then pick the next 5 times at other places when there is a choice of heading left or right. These could be when entering a large public restroom, boarding a train, during two walks in the park, and visiting a seat-yourself restaurant. If you turned the same way 7 or more times out of the 10 total choices, you may want to take note of a space bias.

Now for the results of the star cancellation task. There are 56 total small stars. Count up all the small stars you crossed out on the left side of the page and then again on the right side. Did you get more of them on one side of space? Also look to see if you crossed out the same star more than once and if so, did you do that more often on one side than the other? Do you remember which side you started on or if one side felt like it took longer to find the right stars? Did you feel like you were more distracted by the words or big stars when they were on one side? Some of the questions are easier to answer if you watch a video of yourself, or evaluate someone else, as they are doing the task. If you tend to answer "yes" to these questions, you may have a space bias.

If you have a space bias, you are missing out by not being aware of everything that is going on around you. Ignoring parts of the environment, even a little bit, could also be dangerous. To address this perceptual blind spot, try the following activities. For one day, use only the hand on the side you tend to ignore for as many chores as possible. Try eating, brushing, your teeth, and even writing that way. In addition, be mindful of the foot on the ignored side—dress it first and kick a few things around. Alternatively, do these things for 1 hour per day for 7 days.

Why this helps: It has been found that being unable to use the body on one side leads to ignoring that side of space, similar to what is seen in hemineglect patients. Therefore, strengthening a weaker, less-used, side of the body will conversely increase the ability to pay attention to the space from that side. This may be especially useful for people who are strongly right-handed.

February Take In a Ventriloquist Show

If you have never seen a ventriloquist in action, you are missing a treat. The perceptual illusion is so compelling that it really does seem like the puppet is alive. Do the following and if you can't get to Vegas, try it at the movies or even with your own television. Look at the stage or screen and decide where it sounds like a voice is coming from. It should really sound to you like the puppet is speaking! It's an illusion, of course. Try this, too, to demonstrate to yourself that it is an illusion: When you have a good idea of the location, shut your eyes. Now where does it sound like the voice is coming from? It should be different and now coming merely from the ventriloquist (or stereo speaker). Open your eyes again: Did it jump back to the screen? If the sound seemed to come from different places in the presence of vision and in its absence then you know for sure that you have experienced the illusion of visual capture for yourself, in this case the ventriloquism effect (see Chapter 8). The sound is really coming from the ventriloquist (or the theatre audio speakers) which you can hear just fine when vision is not causing trouble. Be sure to sit close enough to the action so that the two different places—the one indicated by audition and the one indicated by vision—will be noticeably far apart. (If the sound comes from multiple speakers, sit closer to one of them so that they do not cancel.)

Next, try something that is rarely discussed: Try to override the ventriloquism effect by mustering all your powers of conscious reasoning, common sense, and attention. "The sound is not coming from the dummy. The sound is not coming from the dummy. The sound is not coming from the dummy." Try to pay very close attention to where the sound of the voice is actually coming from even with your eyes open. Did it work? If so, it would be an amazing skill to acquire—to influence perception when you want to. It may also make your perception more accurate. Beware though because if you are able to bypass perception's safety nets (see Chapter 8), the world may look more confusing than it does now. Truthful but confusing.

Finally, try your hand—and mouth—at being a ventriloquist yourself. All you have to do is open and close a puppet's mouth at the same time as you speak without moving your mouth. If you can't keep your mouth from moving, hide or cover it. Maybe you can work an Old West bandit with a bandana face mask into your act. Try it out on an audience of friends and see if you can succeed quickly at becoming a ventriloquist without years of training that ventriloquists usually require.

Why this works: Ventriloquism does not involve learning to throw the voice. Instead it is an illusion created by our own perceptual system. You can greatly speed up the acquisition process of becoming a ventriloquist yourself when you know how to fool other people's perception.

March Visualize Your Pain Away

You also don't need years of training with Buddhist monks to use your mind to improve well-being (Chapter 11). We have developed a short 12-minute perceptual training course designed for improvement within just a few sessions. Our subjects report both less stress and less pain. Please note this is not medical advice. Medical advice can be obtained only from your physician.

Readers of this book can access the course for free by entering the password "thatbookwascool" at www.u.arizona.edu/~bedford

Here is a transcribed excerpt from the audio course: "...Now that you're sitting down, look down at the ground beneath you. You will find a magnifying glass. Please pick up the magnifying glass. Has anything on your body or mind been bothering you? Look through the magnifying glass at that part of your body. Look through the glass and note it is getting bigger and bigger and bigger. You can see details of that part now and it takes up all of your vision. Look carefully at what you are seeing. What do you notice? Listen also. Is it telling you anything? Listen hard—is it trying to tell you what is wrong or how to fix it? Go with your gut sense..."

April Test Your DNA

Having your DNA analyzed is an inexpensive and easy to do, even for the squeamish. There is no blood involved and even the driest mouths can part with a little saliva. Think of chocolate cake—yum—to get your juices flowing and spit into a test tube. Or swab the inside of your mouth with a q-tip like brush. In about a month the results will be in and you will know if you have the genes to be a super taster, how well you learn from your mistakes, and how good your verbal memory is. All of these are at least related to perceptual learning (see also Chapter 12). Be warned though: You will also find out if you are at special risk for Alzheimer's and if your ancestry is what you thought it was. Hundreds of thousands of positions along the genome can tell you a lot about yourself.

You may want to search the internet for "genetic test" because new companies for home testing are becoming available more and more frequently. At the time of this writing, www.23andme.com is an excellent choice. They will test nearly one million positions along your genome and show you the results on-line (there are too many to print!) for all sorts of traits, both physical and psychological, in an organized easy-to-read layout. Updates are also posted as new interpretations of the genetic data become available from new research studies. I have no financial interest in the company. Another option is Family Tree DNA (www.ftdna.com), an outstanding leader in genetic ancestry testing for more than a decade. I have no financial interest in FTDNA either. They only provide ancestry information, but all companies let the customers download their DNA data. With the raw data in hand, you can run it through third party software for more analyses. One of my favorite tools for doing this is a free program currently available called *snptips*. It is sitting around unobtrusively in the background as I type this. Any time you come across on the web a position along the genome that you find interesting, the add-on lets you mouse over that position to see your own results.

For example, I came across a summary of a study[2] that claimed that whether or not your parents' educational level influenced how much you learned in school was determined by your genetic variant at position rs1800497 along the genome. I moused over this rs number and saw my result was AG, though I don't know which parent gave me the "A" and which the "G". According to the study, this meant that, yes, there would be a relation between my parents' education and how much I learned in school. No connection was found for those who have instead the variant AA. This position along the genome, rs1800497, affects levels of dopamine, a neurotransmitter involved in how much one is stimulated by the environment (and time perception, see Chapter 10). Having the result "A" means there are fewer dopamine receptor sites, less dopamine, and less central arousal—the person is less stimulated by the environment. Each copy of A also puts one at higher risk for alcoholism and other drug addictions, as it is believed people seek out the missing stimulation through self-medication. I sure wish now I knew which parent I got that "A" from... This example is particularly relevant to perceptual learning because the amount of central arousal may also influence how much change to perception occurs with experience. This means that whether a person has AA, AG, or GG may also influence how much perceptual learning there will be.

Another thing you can do with your raw data is just to look at it in a word processor. When I did this, I was amused to see the number of pages in the document was 16,000. If you were writing a 350 word essay, it would take you 16, 000 of those essays to fill the same space.

Relatively few studies are available for how the genetic results bear on your psychological profile compared to physical traits and medical risks. Fewer still show how genes are relevant to perception or learning. But there are enough to be worth checking out and results of new research studies are coming in all the time. By the time you test, my own results on spatial challenges might be available. Wait until April 25, National DNA Day, to order your own test and you might just get a discount.

May Exercise Your Senses

We hear more about flexing our muscles than we do about flexing our senses. Put on a blindfold and eat an entire meal without vision. How does it taste? Also, when vision is restored, try adding non-toxic red food coloring to apple juice. Assess if it tastes any different just because it looks different. Try it on a friend and see if they can even taste that it's apple juice when the liquid is an unexpected color. Smell every flower on your walk home (and see entry for November "smell like a dog"). Sure, it's a cliché, but you need to exercise those olfactory receptors that are just starving for stimulation. Smell everything your nose can get can get its hands on, though note most people don't take kindly to getting sniffed. Now for audition. Hide a cordless phone, push the locator button on the base, and ask a friend to find the phone using just its sound. Then switch roles. There are two few opportunities to use the sounds that objects make without an assist from another sense modality. To exercise proprioception, mentally trace out the outline of your body in your mind's eye while your eyes are closed (Chapter 3). You may need to wiggle your toes and move a bit to feel every part until you have had more practice. Use your non-dominant hand to brush your hair (not at the same time...). Now with eyes closed again, first extend both arms to the sides and then bring them together so that the tip of each index finger touches the other at about chest height. Practice until you can make them meet quickly and perfectly on the first try of all subsequent attempts. Although vision gets the benefit of frequent practice, there is still plenty of room for improvement. Sit in front of a partially open door and draw it, but try drawing what you actually see, not what you think you should see. Finally, I am not sure what task to recommend for honing the immune system. Perhaps wait for flu season.

Why it helps: The more practice you get with perceptual stimuli from taste, smell, sound, touch, vision, and the immune system, the more your ability to perceive will improve. Isolating a sense modality from the others will enhance attention which is needed to become a perceptual expert (Chapters 6 and 7).

June Be Sure Time is on Your Side

Pick a time when you will be engaging in an activity in one place for at least an hour. Reading a book or preparing dinner would be good choices. Have your trusty assistant set a timer for an unknown amount of time between 1 minute and 1 hour. You can even do this yourself with a digital timer and a little dexterity. Of course, make sure no clock is visible throughout the activity. When the buzzer sounds, write down how much time you think has elapsed. Now check the time and record how much time actually elapsed. How close did you get? Also include a third column briefly describing your activity and how much you enjoyed it, since this can influence your perception of time (Chapter 10). Repeat the challenge at least 4 more times over the next week or so and see if there are any patterns. Do you consistently undershoot and think less time elapses than it really does? Overshoot? Or maybe you have a high variable error and are all over the place with your estimates, one time thinking 23 minutes is 15 minutes and another time thinking it is 50 minutes. Regardless of how good you are to start, by practicing you will get even better. Soon, you will know how much time elapses with far greater precision and accuracy than you did before. Accurate estimation of time is of great practical value, as well as sharpens temporal awareness generally. But note that if you are always wrong in your daily life for a particular interval of time, you may need to specifically practice times close to that duration. Shorter or longer time frames may not generalize to the times you need to improve.

Why this works: You can reduce variable error (the variability in estimations) through the perceptual learning process of perceptual expertise and reduce constant error (consistent over and underestimations) though the perceptual learning process of perceptual adaptation. Just practicing without even knowing if you were right will reduce your variability. Adding the feedback on how close you were to the actual elapsed time will reduce any systematic over or underestimates (Chapter 4).

July Lost in Space

Take our space questionnaire, which is currently at http://www.instant.ly/s/JCa5PABGwAA If the link is broken, send email to me at Bedford@u.arizona.edu. Think you are spatially challenged? (Chapter 9) Email me at the same address. True or false: "I can navigate easily using the directions of north, south, east, and west". That is one of the 37 questions on the survey. We are currently working on a left-right confusion task to also complete on the web.

August **Faces of Another Race**

If your ability to recognize people's faces in either another race or just your own is not as good as you would like it to be (Chapters 5 and 6), there are things you can do. This is going to sound silly but we are finding that it works: Cut out the heads of 10 faces from a group you wish to improve upon and just look at the heads. Attention matters (Chapter 7) and those bodies are distracting, even when you try to attend to just the faces. Look at each disembodied face twice. Now focus on just the eyes. Go through all ten and inspect how each set of eyes differs from the rest. Try to put the photos in piles with similar eyes. Then pick any one of the faces and try putting the remaining 9 in order from least to most similar to the target with respect to the eyes. Look at all 10 pictures again to see if they now look any more different from one another such that you could easily recognize each one without making a mistake.

Still not having much success? Cut out the eyes. You may want to enlarge the images considerably before doing this. Ignore any strange looks (I mean from people not the eyes!) since you will also need to get used to suspicious sideway glances during other activities, such as the blind shopping at the supermarket and playing with body parts (see October and November). Pick out the 2 eye photos from the 10 that are the most different from each other. Look back and forth between them and confirm that they are very different. This step should be easy. Now pick 2 more that are a little more similar to one another, but you can still tell them apart. Look back and forth until it is very obvious to you how different they are. Keep trying 2 pairs of eyes side by side. If you can't tell them apart, try another comparison that looks easier and then try the harder ones again later. Keep it up until every pair of eyes looks different, remembering to spend time on the easier ones first before going on to the harder ones. If you haven't a clue which ones are easier since every one of them looks indistinguishable to you, a friend who is better than average at recognizing faces would be of great help here to presort the eyes for you. Incidentally, you can also use the cutouts of faces and eyes and the comparison study techniques to track down family resemblances. If you have always wondered who you resembled but have been hopeless at such things, you can now acquire the ability to see the resemblance for yourself. Repeat with noses, chins, and overall shape if desired after you're through with eyes, the most important clue for recognizing a familiar face, if you want to see who inherited the family witch chin, Romanesque nose, or full moon face.

Why this works: You can become a perceptual expert at anything, including seeing faces. For some things, we can shave *years* off the time it would ordinarily take to become an expert by following acquisition principles such as attention and starting with easy examples and progressing to the hard ones. See Chapter 7 for all the acquisition principles.

September Know the Body from Which You Perceive

Measure the distance between your shoulders and the distance from shoulder to the tip of outstretched index finger. Divide the shoulder distance in half (divide by 2) and then divide that number by the measured arm length. The average for young American men is .253 and for women, .257. If your ratio is larger, that would mean proportionately wider shoulders than average and if smaller then proportionately longer arms than average. Arm and shoulder dimensions factor into accurate pointing to objects and can predict individual propensity for different sports (and clothes!). Next, measure the distance from ear to ear putting the tape measure around the back of the head. Compare your whole family—how much difference do you find? How would that change your sound localization if you switched heads... Next, are you left or right eye dominant? If you don't know, stand in front of a corner of the room. Extend your arm and index finger and point to the corner (where the two walls and the ceiling all meet). Close each eye in turn and note when the finger appears to jump to the side. If the finger stayed in the same place when just your left eye was open then you rely more on the left eye but if it stayed in the same place for your right eye, it's your right eye that is dominant.

The dominant eye is frequently on the same side as the dominant hand. The eye that is dominant though can change throughout life if one eye becomes weaker, so it's a good idea to recheck every now and then.

Why do this: Basic sensory assessments—there are many you can do—along with DNA analysis (April activity) will give you information on the foundation from which perceptual learning operates (See also Chapters 2 and 9). Besides, we can.

October Excuse Me, I Think That Hand Over There is Mine

You can't let all those body parts available during Halloween go to waste. As you go shopping for a spare, pick an arm that looks human and as like yours as possible. Your task is to actually get that thing to feel like it is *part of you.*

Start by placing your own forearm and hand out of sight. If you can figure out a way to do that while it sits next to the severed arm, that will work better but if not, put your own arm under the table while the severed one is on top. Now have a friend stroke both the fake hand and the real hand for a few minutes, but it has to be at exactly the same time and rate. It won't work, for example, if he uses one of his hands to stroke your real hand and his other hand to stroke the fake hand. Even the most minute deviation from synchrony would clue in your mind that the fake is not yours. Perception may be lazy and only changes when it needs to, but it is also awfully smart. Fashioning a gadget to use on both hands at once can help, such as a large stick that reaches both or the two earpieces on a pair of eyeglasses if the hands are close together. Look at "your hand" on the table—thinking of it as your own won't hurt—as it is being stroked and think about how you can feel your hand being stroked. You can even put your own jewelry on the fake hand before you begin for added realism.

After several minutes, try a few things. One is just to ask yourself: Does it feel like yours? You might not notice that anything is different either way though, even if you really do feel it to be yours now.

Therefore, also have your assistant stop stroking your real hand but keep stroking the fake hand. Did it feel for a split second like your hand was still being stroked? If so, that means you have connected the two. Also try this: Have him take a hammer and hit the FAKE hand. Did you startle? More so than a curious onlooker who was just as close to the hand as you were? If so, it wasn't just loud noises that made you jump but instead was because you sensed that, in some way, *your hand* was getting smashed.

Why it works: The boundaries of our body schema are not fixed. We can gain or lose parts with experience (Chapter 3). Also involved in this rubber hand illusion are the object identity decision (deciding if the stimulation came from one hand or two) and creating a conflict between vision and touch (Chapters 8 and 9).

November Smell Like a Dog!

I do not mean roll around on the ground with dogs and give up bathing. Instead, go to a store that sells cheap spices to stimulate your olfactory sense. Without looking (!), grab 2 at random from each of 3 shelves. I suspect you will get some strange looks as you continue to avoid looking at which spices you are buying all the way through checkout. You better wait until you get home to do any more... Without reading the labels or giving yourself any cues from color or texture, open one jar and take two sniffs. Can you identify it? Either way, wait a few seconds, breath in the next jar, and so on until you have tried all six. Were you able to identify any by name? Or perhaps recognize the scent even if you can't think of the name? Document your experiences and then set about trying to improve them.

To improve identification, now look at the names while smelling them again—remember, putting names on things helps! (Chapter 7.) Try also to attend to what you are smelling, since attention is another principle that was on our list of how to acquire perceptual expertise. Then mix up the spices and take the blind test again. Any better? It may take you several trials of practice, but it is worth the effort.

Here's why: Would you accept being visually blind if there was something you could do about it? I'll assume the answer is no. So why then do we all accept how bad we are at recognizing smells? Sure, we don't have as many olfactory receptors as a dog, but now you know that practice and the other acquisition principles can make you an expert. What kind of new worlds will open up for you if your sense of smell was 10 times better than it is now? Stop letting your dog have all that tree-sniffing fun.

December Make Those Christmas Lights Dance

Many years ago, Marty Seligman and I told a dinner table full of people that we had made the delicious cabbage soup ourselves, when really we had bought it at a local deli. I can't remember exactly why we did that, but there is something appealing about pleasing your friends with hard work that you do for them. And if you can make it appear to be hard work without having to actually do anything, so much the better. You can use your new knowledge of perception to do this and you don't' even have to heat up any soup.

In time for the holidays, dig out that string of wire with the little randomly blinking lights. Play some of your favorite music with a good repetitive beat and watch as the lights look like they are turning on and off in time with the music. It will look like you spent quite a bit of time preparing a coordinated show when, in fact, you did nothing.

Why it works: Vision says the timing of an event is random but audition says it's happening at regular intervals. Audition wins and you perceive the event of the blinking lights as regular. These kinds of conflicts between the senses (Chapter 8) are a prelude to perceptual adaptation (Chapter 9), one of our two major kinds of perceptual learning (Chapter 12).

Will you share your knowledge of perception, or keep it a secret?

I wanted the last chapter to be about something concrete that you could actually do with your knowledge of perceptual learning. This last step is easily forgotten and I have been as much to blame as anyone.

Finally, it is true that I can turn perception every which way, and I even used to turn a few heads, but I cannot turn a phrase. So I end this journey with the words of another[3]:

"It fits in with a French idea of what life is about—the enjoyment of one's senses..."

———

Appendix

The Growth Transformations: Proprioception and Audition

The following steps show how I derived equations (1) and (4) in Chapter 9. These equations involve the type of plasticity needed to accommodate growth of the body to maintain accurate pointing to objects with the arm *(manual growth transformation)* and localizing objects with the ears *(auditory growth transformation)*.

Manual Growth Transformation

The manual growth transformation characterizes the error that will be made in pointing following growth and before adaptation. It is equivalently a description of the type of plasticity that would be needed to accommodate growth of the shoulders and arms. Finally, it can also be viewed as characterizing the disagreement between vision (assumed unchanged by growth) and proprioception following growth as a function of where the target is in space.

Consider a distant target that is presented to the right of straight-ahead to which the observer must point. She attempts to point to the target with her right hand, but is not permitted visual feedback of any part of the arm or hand. After growth and before adaptation, the child is calibrated for a smaller-sized body and therefore, feeling her arm to be where it was before growth, directs the arm with the same shoulder angle that would have produced accurate pointing when she was smaller. With shoulder width growth, the shoulder and arm are further to the right of where they were before. Thus, the arm will be in a position further to the right than where she feels it and she will overshoot the target. Arm length growth has the opposite effect and she will undershoot the target.

Figure A2-1 shows a schematic of this pointing situation before and after

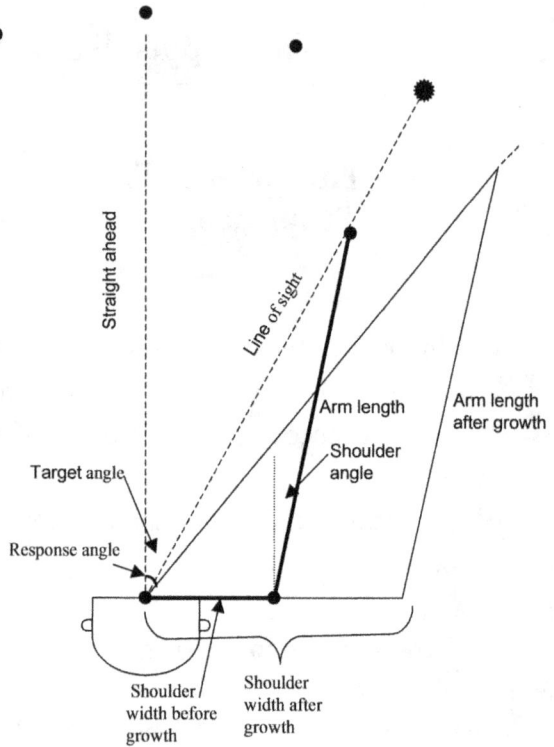

Figure A2-1 Manual pointing to a distant target (larger sunburst) before and after growth

growth. The intended target is shown as a larger circle. When the head and trunk are aligned, the straight-ahead direction from the center of the head, or cyclopean eye, is the line perpendicular to the trunk and is shown as a dotted line. The line of sight from the cyclopean eye to the target can also be identified and also shown as a dotted line. Pointing in this context refers to rotating the entire arm from just the shoulder joint such that the eye, tip of the finger, and the visual target in space are all aligned. When the observer attempts to point to the target with an outstretched right arm, the tip of the index finger on the right hand would need to intersect the line of sight from the eye to the target to accurately match the (azimuth) angle of the target. Note how the tip of the (index finger on the) arm before growth, shown with a bold line, intersects the line of sight. The angle the observer points to before growth is the angle formed by the straight-ahead line and the line of sight to the target (*target angle*, assuming accuracy). The figure also shows the position where the arm will arrive after growth when the child believes she is pointing to the target, but is no longer accurate. Connecting the line between the cyclopean eye and the new after-growth position of the tip of index finger shows the angle formed by this line and the straight-ahead line, that is, the angle the child is pointing to after growth (*response angle*).

Definitions and assumptions

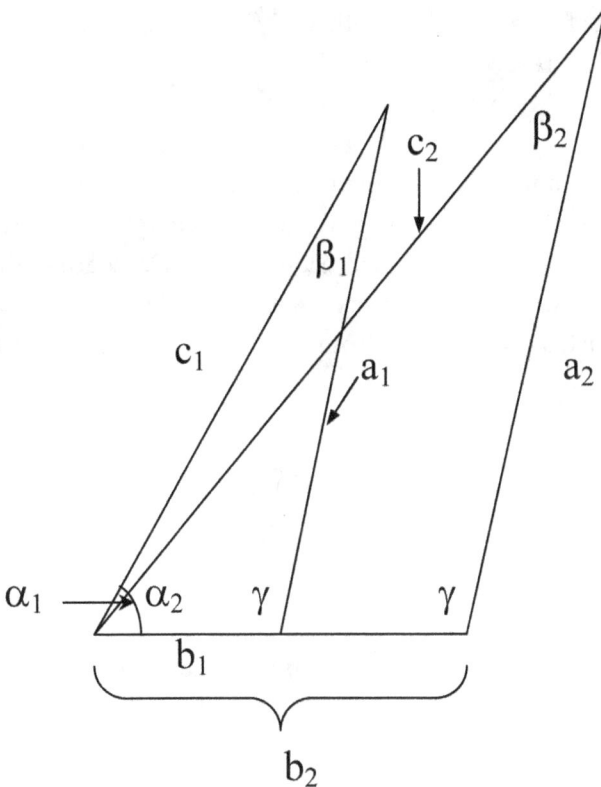

Figure A2-2 The two triangles extracted from the previous figure reflecting before growth pointing with triangle 1 (a1-b1-c1) and after growth pointing with triangle 2 (a2-b2-c2). See text for interpretations of the symbols.

The two triangles from the figure, Triangle 1 (before growth) and Triangle 2 (after growth) are extracted and shown in Figure A2-2. Triangle 1 is defined by sides a1, b1 and c1, and triangle 2 by sides a2, b2, and c2. Side *b1*

= initial width of shoulder from middle of the neck/back to joint (acromial) before growth, **b2** = final width of shoulder after growth, **a1** = initial length of outstretched arm from shoulder joint to tip of index finger before growth, **a2** = final length of arm after growth, **α1** = angle opposite side a1 (= 90 - α1', where α1' = target angle as described in the main text), **α2** = angle opposite side a2 (= 90 - α2', where α2'= response angle as described in the text), **λ** = shoulder angle formed by b1, and a1 = shoulder angle formed by b2 and a2. In addition, **c1** and **c2** are the sides opposite angle λ in triangles 1 and 2 respectively, and represent distances from cyclopean eye to fingertip before and after growth, and **β1** and **β2** are the angles opposite sides b1 and b2 respectively.

Make the additional assumptions that both a1 > b1 and a2 > b2; i.e., arm length is greater than shoulder width before and after growth and then that angle γ is the same in both triangles since the arm will be directed at the same angle as discussed in the main text. Angle α1 is a known quantity when the target angle is specified. Sides a1, b1, a2, b2 are body segments that can be measured from observers. Thus, angle α2 can be determined from α1, a1, b1, a2, and b2, and from α2 calculate α2', the desired response angle after growth.

Derivation

The goal is to derive:

$$\alpha 2 = f(\alpha 1) \tag{A1}$$

From Triangle 2, since two sides, a2 and b2, and the inclusive angle, γ, are known, then from trigonometric identities, we know that the angle opposite a2 is:

$$\tan \alpha 2 = \frac{a2 \sin \gamma}{b2 - a2 \cos \gamma} \tag{A2}$$

Angle γ can be calculated from Triangle 1. In Triangle 1, two sides, a1 and b1, and here the *non*-inclusive angle α1, are known. The sum of the 3 angles is 180°:

$$\gamma = 180 - (\alpha 1 + \beta 1) \tag{A3}$$

and from the law of sines:

$$\sin \beta 1 = \frac{b1 \sin \alpha 1}{a1} \tag{A4}$$

To solve for $\cos \gamma$ in Eq. (A2) substitute the expression for γ from Eq. (3) and expand $\cos \gamma$ to $-(\cos \alpha 1 \cos \beta 1 - \sin \alpha 1 \sin \beta 1)$, using the general expansion for $\cos (X + Y)$ for any angles X and Y, along with $\sin (180) = 0$ and $\cos (180) = -1$. This expression can be further manipulated by substituting $\sin \beta 1$ with the equation for $\sin \beta 1$ given in (4). To solve for $\sin \gamma$ in Eq. (A2), similar expansion of $\sin (X + Y)$ can be used or more directly with $\sin \gamma = c1 \sin \alpha 1/a1$ (from the law of sines) and the identity $c1 = a1 \cos\beta + b1 \cos\alpha$. Substituting the expressions for $\cos \gamma$ and $\sin \gamma$ into Eq. (A2) yields:

$$\tan \alpha 2 = \frac{a2 \sin \alpha 1 (\cos \beta 1 + b1/a1 \cos \alpha 1)}{b2 + a2 (\cos \alpha 1 \cos \beta 1 - b1/a1 \sin^2 \alpha 1)} \tag{A5}$$

All that remains is to express $\cos \beta 1$ in terms of $\alpha 1$. We know from Eq. (A4) that the sine of $\beta 1$ is $(b1\sin \alpha 1)/a1$. One can generate a right triangle whose hypotenuse is a1 and whose side opposite $\beta 1$ is b1 $\sin \alpha 1$. The side adjacent to $\beta 1$ then is the square root of: $a1^2 - b1^2 \sin^2\alpha 1$ (sum of the squares of two sides of a right triangle equal the square of the hypotenuse). $\cos \beta$ is the adjacent side divided by the hypotenuse or square root of: $(a1^2 - b1^2 \sin^2\alpha 1)/a1)$, which is equivalent to square root of: $1- (b1/a1)^2 \sin^2 \alpha 1$. Let $R1 = b1/a1$, the ratio between the shoulder width and the arm length before growth; substituting these in Eq. (A5) to solve for $\alpha 2$ yields:

$$\alpha 2 = \tan^{-1} \left[\frac{a^2 \sin \alpha 1 \left(\sqrt{1 - R1^2 \sin^2 \alpha 1} \right) + R1 \cos \alpha 1}{b2 + a2 \left(\cos \alpha 1 \sqrt{1 - R1^2 \sin^2 \alpha 1} - R1 \sin^2 \alpha 1 \right)} \right] \tag{A6}$$

To more easily see the ratio after growth, multiply the dominator by a2/a2, which rephrases the denominator with b2/a2, the ratio after growth (R2), as the first term. In addition, to make the denominator more comparable to the numerator, substitute $\sin^2\alpha 1$ with the equivalent expression $1 - \cos^2 \alpha 1$

(since $\sin^2\alpha + \cos^2\alpha = 1$). Expanding and pulling out the common elements yields:

$$\alpha2 = \tan^{-1}\left[\frac{\sin\alpha1\left(\sqrt{1-R1^2\sin^2\alpha1}\right)+R1\cos\alpha1}{R2-R1+\cos\alpha1\left(\sqrt{1-R1^2\sin^2\alpha1}\right)+R1\cos\alpha1}\right] \quad (A7)$$

This is equation 1 presented in Chapter 9. Or equivalently:

$$\alpha2 = \tan^{-1}\left(\frac{(\sin\alpha1)\,M}{R2-R1+(\cos\alpha1)\,M}\right) \quad (A8)$$

where $M = \sqrt{1-R1^2\sin^2\alpha1}+R1\cos\alpha1$

Auditory Growth Transformation

To derive the equation that generates auditory localization errors as a function of target position after growth and before adaptation, I began with Woodworth's classic formula: ITD = r/(speed for sound (θ + sin θ)) where r is the radius of the head and θ is the angle of the sound (in radians). This calculates the interaural timing differences (ITD) as a function of head size and angular position of sound (see Note 2 for Chapter 2). I Thus, if we consider the ITD produced by a sound before head growth then:

$$ITD1 = \frac{r1}{\text{speed of sound}(\theta1 + \sin\theta1)} \quad (A9)$$

where r1 is the radius of the head before growth and $\theta1$ is the angle of the sound that would produce that ITD (ITD1). When the child grows to a larger head size, the distances between the ears, and hence the radius of the head, will be larger. This will produce a different set of interaural timing differences to the sounds:

$$ITD2 = \frac{r2}{\text{speed of sound}(\theta2 + \sin\theta2)} \quad (A10)$$

where r2 is the radius of the head after growth and $\theta2$ the angle of the sound that would produce ITD2.

Now consider if the child grows from head size r1 to r2, what the effects of a sound presented at position $\theta2$ would be before any plasticity/adaptation. Since the size of the head initially is the larger after-growth size, a sound at $\theta2$ will produce the interaural timing difference, ITD2, given in equation A10. However, the child's system does not yet know this, takes this ITD2, and uses the smaller scale pre-growth kid-sized relation between sound position and ITD given in equation A9 to infer a location of the sound, $\theta1$, on the basis of the ITD it receives, ITD1 (or equivalently ITD2 in this case). Thus, we can calculate $\theta1$, which is the perceived angle, as a function of $\theta2$, which is the actual angle of the sound, $\theta1 = f(\theta2)$, by combing equations A9 and A10. From equation A9, $\theta1 + \sin \theta1 =$ (speed of sound * ITD1) / R1. We can substitute ITD1 with the expression given in equation A10, since ITD 1 = ITD 2. The speed of sound cancels, leaving:

$$\theta1 + \sin \theta1 = \frac{r2}{r1} (\theta2 + \sin \theta2) \tag{A11}$$

This is the auditory growth transformation shown in the equation 4 of Chapter 9. The value r2 is the radius of the head after growth, r1 the radius of the head before growth, $\theta2$ the angle of the sound, and $\theta1$ the perceived angle of the sound. The auditory growth transformation is comparable to the manual growth transformation in that it characterizes the perceived (response) target location following uncompensated growth as a function of the actual target position.

Notes

Preface

1. Alan Alda's speech to graduates at Rockefeller Center. Alda, A. (2007). *Things I Overheard While Talking to Myself* (1st ed). Random House.
2. The phrase was popularized by the very serious plain-cloths detective Sergeant Friday in the classic TV show *Dragnet*, which aired in the 1950s.
3. From the song *A-Tisket, A-Tasket* recorded by Ella Fitzgerald in 1938.

Chapter 1

1. Rozin, P. & Schull, J. (1988). The adaptive-evolutionary point of view in experimental psychology. In R. C. Atkinson, R. J. Herrnstein, G. Lindzey, & R. D. Luce (Eds.) *Handbook of Experimental Psychology* (pp. 503-546). New York: Wiley Interscience. *See also* Bedford, F. (1993.) Perceptual Learning. In D. Medin (Ed.) *The Psychology of Learning and Motivation, 30*, 1-60.
2. Bedford, F. L. (1995). Constraints on perceptual learning: Objects and dimensions. *Cognition, 54*, 253-297.
3. Beeli, G., Esslen, M., & Jäncke, L. (2005). Synaesthesia: when coloured sounds taste sweet. *Nature, 434*(7029), 38. doi:10.1038/434038a
4. See Shallice T. (1988). *From neuropsychology to mental structure.* Cambridge, UK: Cambridge University Press. *See also* Bedford, F. L. (2003). More on the Not-the-Liver Fallacy: Medical, neuropsychological, and perceptual dissociations. *Cortex*, 39, 170-173.

* Note from Table 1-1: "The man who fell out of bed" is a chapter in: Sacks, O. (1985). *Man Who Mistook His Wife for a Hat, and Other Clinical Tales.* Summit Books. It describes a patient who felt as if his leg belonged to someone else. He would push it out of bed and would end up on the floor with his leg.

** Note from Figure 1-1. The discovery of the hidden anatomy was made relatively recently by Frank Meshberger, a gynecologist. Meshberger, F. L. (1990). An Interpretation of Michelangelo's Creation of Adam Based on Neuroanatomy. *JAMA: The Journal of the American Medical Association, 264*(14), 1837–1841. doi:10.1001/ jama.1990.03450140059034

*** Note from Figure 1-2. It is uncertain where to place fatigue. Fatigue does not directly make you function better but if fatigue occurs to prevent injury, it may not belong in the "damage" half of the tree. See also Chapter 12.

Chapter 2

1. Despite a rumor and Timmy's frequent mishaps, Timmy was never trapped in a well in the classic television show *Lassie*, which ran from 1954-1974.
2. See Woodworth, R. S. & Schlosberg, H., (1954). *Experimental Psychology (Rev. ed)*. Holt: Oxford, England
3. Bedford, F. L. (2002). Generality, mathematical elegance, and evolution of numerical/object identity. *Brain and Behavioral Sciences.* Special issue on the work of Roger Shepard, *24,* 664-665.
4. The numbers in the table are not identical to the graph owing to slightly different calculations.
5. See Clifton, R.K., Gwiazda, J., Bauer, J.A., Clarkson, M. G. & Held, R. M. (1988). Growth in head size during infancy: Implications for sound localization. *Developmental Psychology, 24,* 477-483 *and* Fernald, A. (2007). Hearing, listening and understanding: Auditory development in infancy in Bremner, G. and Fogel, A. (Eds.) *Blackwell Handbook of Infant Development.* Blackwell Publishing Ltd: Oxford, UK. pp. 35 -70. doi: 10.1002/9780470996348.
6. The Clifton reference from the previous note.
7. Banks, M.S. (1988). Visual recalibration and the development of contrast and optical flow perception. In Albert Yonas (Ed.), *Perceptual development in infancy.* Hillsdale, NJ: Lawrence Erlbaum Associates, Inc. pp. 145-196. The quoted phrases are from page 149.
8. One perceptual interpretation if the ITD were too large for a sound source is that there really are two sound sources. If some other information indicates that there is only one source then the auditory information provides an internal contradiction. The question as two whether two sensory samples indicate one or two sources is a general challenge in perception—and one of my favorites. I discuss it further in Chapters 8 and 9 (*Object Identity* decision).
9. In *Big*, the Tom Hank's character turned from a small pre-adolescent boy to a large full-grown man overnight after wishing to a fortune telling machine. In *Dark Shadows*, an infant had accelerated growth to a toddler, a boy, and then a man, all over a period of days. The soap opera *Passions* showed Tabitha's witchlette daughter Endora turn instantly from a toddler to a young woman when she chose.

Chapter 3

1. Rowling, J. K. (2007). *Harry Potter and the Deathly Hallows* (1st US ed). Arthur A. Levine Books. See page 749.
2. Collins, S. (2009). *Catching Fire. Book 2 of the Hunger Games Trilogy.* Scholastic Press. See page 289.
3. Maravita, A., & Iriki, A. (2004). Tools for the body (schema). *Trends in Cognitive Sciences, 8*(2), 79–86. doi:10.1016/j.tics. 2003.12.008

4.   Bedford, F. L. (2007). Is prism adaptation "for" growth? *Perceptual and Motor Skills*, 105, 351-354.

5.   Ehrsson, H. H. (2007). The experimental induction of out-of-body experiences. *Science (New York, N.Y.), 317*(5841), 1048. doi:10.1126/science.1142175 *and* Lenggenhager, B., Tadi, T., Metzinger, T., & Blanke, O. (2007). Video ergo sum: manipulating bodily self-consciousness. *Science (New York, N.Y.), 317*(5841), 1096–1099. doi:10.1126/ science.1143439

6.   Kubovy, M. (1988). *The Psychology of Perspective and Renaissance Art.* Cambridge University Press.

Chapter 4

1.   Wagman, J. B., Shockley, K., Riley, M. A., & Turvey, M. T. (2001). Attunement, calibration, and exploration in fast haptic perceptual learning. *Journal of motor behavior, 33*(4), 323–327. doi:10.1080/00222890109601917 The quote from the procedure section is on page 325.

2.   Shepard, R. N. (2001) Perceptual cognitive universals as reflections of the world. *Brain and Behavioral Sciences, Special issue on the work of Roger Shepard, 24*, 581-601.

3.   Gibson, E. J. (1953). Improvement in perceptual judgments as a function of controlled practice or training. *Psychological Bulletin, 50*(6), 401–431. doi:10.1037/h0055517

4.   The answer to the questions are as follows. Before training: CE=.2, VE=.255, AE= .25; After training: CE=.2, VE=.0, AE= .2; No, the training did not involve feedback because constant error (CE) did not decrease. Feedback is needed to change the constant error. Variable error was completely eliminated, but that can occur without feedback.

Chapter 5

1.   Ardmore, PA regional rail, March 2007. If you're still out there, Mr. Taxi Cab Driver, thanks.

2.   Doty, N.D. (1998).The influence of nationality on the accuracy of face and voice recognition. *American Journal of Psychology, 111*, 191-214.

3.   Platz, S. J., & Hosch, H. M. (1988). Crossracial/ethnic eyewitness identification: A field study. *Journal of Applied Social Psychology*, 18 (11), 972-984.

4.   We have though found little of an Other Race Effect between Caucasian and Latino college students in Arizona. Our Latino subjects have been especially good on all of our face perception tasks whether involving race or not, an unexpected finding we hope to be able to follow-up on.

5.   Deregowski, J. B., Ellis, H. D., & Shepherd, J. W. (1975). Descriptions of White and Black faces by White and Black subjects. *International Journal of Psychology, 10*(2), 119–123.

6. See the dissertation of my former graduate student for this finding as well as for an early review of neural mechanisms: Reinke, K. S. (1998). *Visual and neural plasticity: A study of line orientation discrimination* (Ph.D.). The University of Arizona, United States -- Arizona. *Dissertation Abstracts International - B 59/11*, p. 6088, May 1999. *See also* Shiu, L. P., & Pashler, H. (1992). Improvement in line orientation discrimination is retinally local but dependent on cognitive set. *Perception & psychophysics, 52*(5), 582–588.

7. What does doing worse than baseline mean? It means if we could find people who are not experts on any race, then those who are experts on one race would do worse on other races than these novices. This prediction is indeed confirmed with infants—the ultimate novices—who do better at other-race faces than adults do!

8. Penny demo from cognitive psychologist Henry Gleitman at the University of Pennsylvania.

9. Arizona v. Youngblood is a famous case known nationally, but not because of cross-racial identification problems. The case made it all the way to the US Supreme Court because of evidence that had not been refrigerated for potential later analysis. The Supreme Court ruled that a defendant is not denied due process when there is a failure to preserve evidence, provided there was no malicious intent involved in its destruction. The case made headlines again when the DNA evidence proved that Youngblood was not the rapist. Sources include Kent Cattani, Chief Counsel, Capital Litigation Section in a memo to Terry Goddard, Attorney General (2005) and Gross et al.

10. After too much thought, no doubt touching my chin all the while, I concluded that chin-touching really just goes along with thinking. When does law enforcement see people think? Sadly, too often when someone is trying to invent a story about how things could have happened!

11. Thompson P. (1980). Margaret Thatcher: A new illusion. *Perception, 9*, 483-48.

* Note from Figure 14: Bugeleski and Alampay's original experiment showed subjects either pictures of animals or people to influence what they saw in the ambiguous rat-man figure. They say in a footnote: "The figure was originally prepared to illustrate the bias of some psychologists who view men as rats or vice-versa". Henry Gleitman popularized a variant demonstration where the rat-man figure was transformed to a clear unambiguous rat or man image that was viewed first. See an early edition of Gleitman, H. *Psychology*. Norton.

Chapter 6

1. Answer to question on gender in the photo: The photo is of a woman.

2. O'Toole, A.J., Peterson, J. & Deffenbacher, K.A. (1996) An 'other-race effect' for categorizing faces by sex, *Perception, 25*, 669-675. Quote from page 670.

3. Why doesn't the fact that the Japanese pictures are harder also imply that

Japanese faces are intrinsically harder to tell apart than Caucasian faces? Making a small change to the picture set reverses the relative difficulty of the picture sets. O'Toole and colleagues removed the exterior hair shape because before they did that, it was the Japanese pictures that were easier for all races rather than the Caucasian pictures. The authors believe gender-specific hairstyles provided more of a cue to gender for Japanese pictures and gender-specific eyebrow shaping was more of a cue to gender for American pictures.

4.   Gibson, J.J. & Gibson, E. (1955). Perceptual learning: differentiation or enrichment? *Psychological Review, 62*, 32-41. The quote is from page 38, italics added.

5.   www.u.arizona.edu/~bedford. Follow links to the book and you may need to enter password: Iamlearningalot. The positions of the target stimuli (correct answer "yes") for the experiment are 15, 16, 19 and 23. Begin counting from the first unnumbered picture. On each page, continue with the first item on the page. When there is more than one picture on a page, precede clockwise zig-zagging back and forth. There are 33 pictures total.

6.   To come full circle, why not use the Gibsons' paradigm that illustrates perceptual expertise to investigate the Other Race Effect? It belongs in this spot, but I have already squeezed in more than a long tedious chapter's worth of answers, tasks, graphs, data, and details. This study will be described towards the end of the next chapter.

Chapter 7

1.   Gibson, E. (1969). *Principles of Perceptual Learning and Development.* Prentice Hall College Division.

2.   Ahissar, M. (1999). Perceptual learning. *Current Directions in Psychological Science, 8*,124-128.

3.   Biederman, I. & Shiffrar, M.M. (1987) Sexing day-old chicks: A case study and expert systems analysis of a difficult perceptual-learning task. *Journal of Experimental Psychology: Learning, Memory, and Cognition,13(4)*, 640-645. Biederman & Shiffrar used pictures to illustrate instructions, which likely works faster than if one were to just use words. The answers to the sex of the chicks in our figure, going from left to right, are: female, male, female. (No clue as to the chick with egg on the bottom of the figure.)

4.   See Note 6, Chapter 5.

5.   See e.g., Markman, E.M. (1990). Constraints children place on word meanings. *Cognitive Science, 14*, 57-77.

6.   Neisser, U., & Becklen, R. (1975). Selective looking: Attending to visually specified events. *Cognitive Psychology, 7*(4), 480–494. doi:10.1016/0010-0285(75)90019-5 *and* Chabris, C., & Simons, D. (2010). *The Invisible Gorilla: And Other Ways Our Intuitions Deceive Us* (1st ed.). Crown Archetype.

7.   See Leekam, S. R., Nieto, C., Libby, S. J., Wing, L., & Gould, J. (2007). Describ-

ing the sensory abnormalities of children and adults with autism. *Journal of autism and developmental disorders*, *37*(5), 894–910. doi:10.1007/s10803-006-0218-7

8. See Wolfe, J. M., Kluender, K. R., & Levi, D. M. (2011). *Sensation & Perception, Third Edition* (3rd ed). Sinauer Associates, Inc.

9. Feingold, G. A. (1914). The influence of environment on the identification of persons and things. *Journal of Criminal Law and Political Science, 5*, 39–51.

10. Bedford, F. L. & Panagos, M. (May 2005). American Psychological Society, Los Angeles, CA. *The illusory misidentification of human faces: Implications for eyewitness accounts, and face perception and memory.*

*Answers from Figure 7-9: One trash can instead of two, shadow of raised hand missing (3rd from left in the front row), two students missing (last row at the right), an extra palm tree on the left, T-shirt of student in front row no longer has a logo (3rd from left). Bonus item: Dr. Bedford's water bottle is now a can of Coke!

Chapter 8

1. Botvinick, M., & Cohen, J. (1998). Rubber hands "feel" touch that eyes see. *Nature, 391*(6669), 756. doi:10.1038/35784

2. Rock, I., & Victor, J. (1964). Vision and Touch: An Experimentally Created Conflict between the Two Senses. *Science, 143(3606)*, 594-596.

3. See Note 4, Chapter 3

4. See Note 6, Chapter 3

5. Jackson, C. V. (1953). Visual factors in auditory localization. The *Quarterly Journal of Experimental Psychology*, 5, 52–65.

6. Bedford, F. L. (2001). Towards a general law of numerical/object identity. *Cahiers de Psychologie Cognitives/ Current Psychology of Cognition, 20,* 113-175; Bedford, F. L. (2001). Object Identity Theory and the nature of general laws. *Cahiers de Psychologie Cognitives/ Current Psychology of Cognition, 20,* 277-293; Bedford, F. L. (2001). The role of object identity and Klein's geometry in cross-modal and other discrepancies. *Cahiers de Psychologie Cognitives/ Current Psychology of Cognition, 5,* 381-394 *and* note 2, Chapter 2.

7. See also *modality precision hypothesis* for an early related discussion in Welch, R. B. (1978). *Perceptual Modification: Adapting to Altered Sensory Environments.* Academic Press, Inc. And for some of the complexities involved in modality precision and "dominance", see van Beers, R. J., Wolpert, D. M., & Haggard, P. (2002). When feeling is more important than seeing in sensorimotor adaptation. *Current biology: CB, 12*(10), 834–837.

8. Bertelson, P., & Radeau, M. (1981). Cross-modal bias and perceptual fusion with auditory-visual spatial discordance. *Perception & Psychophysics, 29*(6), 578–584. doi:10.3758/ BF03207374. *See also* Bertelson, P. (1998). Starting from the ventriloquist: The perception of multimodal events. *Advances in*

Psychological Science, Vol. 2: Biological and Cognitive Aspects, pp. 419–439. Psychology Press. For a comment on Radeau and Bertelson's analysis of the ventriloquism effect, see Bedford, F. L. (1994). A pair of paradoxes and the perceptual pairing process. *Cahiers de Psychologie Cognitvies/Current Psychology of Cognition, 13,* 60-68.

9. Mishkin, M., Ungerleider, L.G., Macko, K.A., 1983. Object vision and spatial vision: two cortical pathways. *Trends in Neurosciences 6,* 414–417; Goodale, M.A. & Milner, A.D., 1992. Separate visual pathways for perception and action. *Trends in Neurosciences, 15(1)* 20-25; Bridgemen, B. (1993). Separate visual representations for perception and for visually guided behavior. In S. R. Ellis, M. K. Kaiser, & A. C. Grunwald (Eds.), *Pictorial communication in virtual and real environments,* pp. 316-327. Philadelphia, PA US: Taylor & Francis *and* Dassonville, P. & Bala, J. K. (2004). Perception, action, and Roelofs effect: a mere illusion of dissociation, *PLoS Bio.* 2, e364.

10. Bedford, F. L. (2007). Can a space-perception conflict be solved using three sense modalities? *Perception, 36, 508-515.*

Chapter 9

1. Kohler, I. (1962). *Experiments with Goggles.* W.H. Freeman Company *and* Kohler, I. (1964). The Formation and Transformation of the Perceptual World. *Psychological Issues, Monograph,* 1-173. For the earliest work on adaptation to turning the world upside down, *see* Stratton, G.M. (1897), Vision without inversion of the retinal image. Psychological Review, 4, 341-360, 463-481.

2. Bedford, F. L. (1999). Keeping perception accurate. *Trends in Cognitive Sciences, 3,* 4-12. *See also* Bedford, F. L. (2004) Analysis of a constraint on perception, cognition, and development: One object, one place, one time. *Journal of Experimental Psychology, Human Perception and Performance, 30,* 907-912.

3. Bedford, F. L. (1993). Perceptual and Cognitive Spatial Learning. Journal of *Experimental Psychology: Human Perception and Performance, 19,* 517-530; Bedford, F. (1993) Perceptual Learning. In D. Medin (Ed.) *The Psychology of Learning and Motivation, 30,* 1-60 *and* Bedford, F. L. (1989) Constraints on learning new mappings between perceptual dimensions. *Journal of Experimental Psychology: Human Perception and Performance, 15,* 232-248.

4. Bedford, F. L. (1994). Of computer mice and men. *Cahiers de Psychologie Cognitvies/ Current Psychology of Cognition, 13,* 405-426.

5. Harris, C. S. (1965). Perceptual adaptation to inverted, reversed, and displaced vision. *Psychological Review, 72(6),* 419–444. doi:10.1037/h0022616

6. Redding, G. M., & Wallace, B. (1990). Effects on prism adaptation of duration and timing of visual feedback during pointing. *Journal of Motor Behavior, 22(2),* 209–224 *and* Redding, G. M., & Wallace, B. (1997). *Adaptive Spatial*

Alignment (illustrated ed). Psychology Press. For early work on terminal and concurrent pointing, *see* Welch, R. B. (1978). *Perceptual Modification: Adapting to Altered Sensory Environments.* Academic Press Inc.

7.   Harris, C. S. (1974). Beware of the straight-ahead shift--a nonperceptual change in experiments on adaptation to displaced vision. *Perception, 3*(4), 461–476.

8.   Rossetti Y., Rode G., Pisella L., Farne A., Li L., Boisson D. Prism adaptation to a rightward optical deviation rehabilitates left hemispatial neglect. *Nature.* *1998,* 395, 166-166-169 *and* Rossetti, Y., Jacquin-Courtois, S., Rode, G., Ota, H., Michel, C., & Boisson, D. (2004). Does action make the link between number and space representation? Visuo-manual adaptation improves number bisection in unilateral neglect. *Psychological Science, 15*(6), 426–430. doi:10.1111/j.0956-7976.2004. 00696.x

9.   Held, R. (1965). Plasticity in sensory-motor systems. *Scientific American, 213*(5), 84–94 *and* Held, R., & Hein, A. (1963). Movement-produced stimulation in the development of visually guided behavior. *Journal of Comparative and Physiological Psychology, 56*(5), 872–876. doi:10.1037/h0040546

10.   Simmons, K. (1944). The Brush Foundation study of child growth and development II. Physical growth and development. *Monographs of the Society for Research in Child Development, 9*, 1-86.

11.   Ferrel, C., Bard, C., & Fleury, M. (2001). Coordination in childhood: modifications of visuomotor representations in 6-to 11-year-old children. *Experimental Brain Research*, 138, 313-321

12.   See the Bedford reference in Note 4, Chapter 3.

13.   To read about how blind children perform at this task, see Landau, B., Gleitman, H., & Spelke, E. (1981). Spatial knowledge and geometric representation in a child blind from birth. *Science, 213*(4513), 1275–1278.

Chapter 10

1.   Cunningham, D.W., Billock V.A. & Tsou, B. H. (2001). Sensorimotor Adaptation to Violations of Temporal Contiguity. *Psychological Science 12(6,)* 532-535. The quote on page 251 in Chapter 10 is from page 553 in the article.

2.   For object identity applied to time, specifically transformation geometry, see note 6 for Chapter 8.

3.   Held, R. & Durlach, R. (1993). Telepresence, Time Delay and Adaptation in Stephen R. Ellis (Ed.), *Pictorial Communication in Virtual and Real Environments*, pp 232-246, London, UK: Taylor and Francis.

Chapter 11

1.   Ball, T. M., Shapiro, D. E., Monheim, C. J., & Weydert, J. A. (2003). A pilot study of the use of guided imagery for the treatment of recurrent abdominal pain in children. *Clinical Pediatrics, 42*(6), 527–532. doi:10.1177/

000992280304200607

2. Shepard, R & Metzler, J. (1971) Mental rotation of three dimensional objects. *Science, 171,* 701-3.

3. Kabat-Zinn, J. (2005). *Full Catastrophe Living: Using the Wisdom of Your Body and Mind to Face Stress, Pain, and Illness: Fifteenth Anniversary Edition.* Delta (published 1990, Random House)

4. Bedford, F. L. (2012). A perception theory in mind-body medicine: guided visual imagery and mindful meditation as cross-modal adaptation. *Psychonomic Bulletin and Review, 19, 1,* 24-45. DOI: 10.3758/s13423-011-0166-x

5. Blalock, J. E. (2005). The immune system as the sixth sense. *Journal of Internal Medicine, 257*(2), 126–138. doi:10.1111/ j.1365-2796.2004.01441.x

6. Spanos, N. P., Williams, V., & Gwynn, M. I. (1990). Effects of hypnotic, placebo, and salicylic acid treatments on wart regression. *Psychosomatic Medicine, 52*(1), 109–114.

7. Collins, M. P., & Dunn, L. F. (2005). The effects of meditation and visual imagery on an immune system disorder: dermatomyositis. *Journal of Alternative and Complementary Medicine), 11*(2), 275–284. doi:10.1089/acm. 2005.11.275. Note that transcendental meditation involves completely clearing the mind rather than the ultra attention of mindfulness meditation. This, too, appears to require unusual attention skills, in this case attending to "nothing".

8. Editors, P. M. (2003). *The Doctors Book of Home Remedies Revised Edition.* Bantam.

9. Lane, R. P., Roach, J. C., Lee, I. Y., Boysen, C., Smit, A., Trask, B. J., & Hood, L. (2002). Genomic Analysis of the Olfactory Receptor Region of the Mouse and Human T-Cell Receptor α/δ Loci. *Genome Research, 12*(1), 81–87. doi:10.1101/ gr.197901

10. Nunes, D. F. T., Rodriguez, A. L., da Silva Hoffmann, F., Luz, C., Filho, A. P. F. B., Muller, M. C., & Bauer, M. E. (2007). Relaxation and guided imagery program in patients with breast cancer undergoing radiotherapy is not associated with neuroimmunomodulatory effects. *Journal of Psychosomatic Research, 63*(6), 647–655. doi:10.1016/ j.jpsychores. 2007.07.004

11. Bedford, F. L. (2011). The missing sense modality: The immune system. *Perception, 40,* 1265-1267. (DOI:10.1068/ p7119).

12. Bedford, F.L. & Peterson, S. *Effect of visual imagery and perceptual mindfulness on physical healing.* Research presented at Association for Psychological Science (APS) conference, May 2010.

Chapter 12

1. Adapted from a line in the poem by Phillip James Bailey, "We should count time by heart-throbs".

2. For example, Welch (see reference in Note 7, Chapter 8) characterized prism

adaptation as "quite a forgettable experience", that it showed decay, and described adaptation as a "semipermanent change"

3. For a Pavlovian conditioning account of the McCollough Effect see Siegel, S., & Allan, L.G. (I987). Contingency and the McCollough effect. *Perception and Psychophysics, 42,* 281-283. For my Perceptual Learning Theory of the McCollough Effect, see Bedford, F. L. (1995) Constraints on perceptual learning: Objects and dimensions. *Cognition, 54,* 253-297; Bedford, F. L. (1997). Are long-term changes to perception explained by Pavlovian associations or Perceptual Learning Theory? *Cognition, 64,* 223-230; Bedford, F.L. (2011) Mystery of the anti-McCollough Effect. *Attention, Perception, and Psychophysics,* 73, 2197-202. DOI: 10.3758/s13414-011-0163-1 *and* Bedford, F. L. & Reinke, K. S. (1993) The McCollough Effect: Dissociating retinal from spatial coordinates. *Perception and Psychophysics, 54,* 515-526.

4. Bedford, F. L. (1995). Localizing the spatial localization system: Helmholtz or Gibson? *Psychological Science, 6,* 387-388.

5. Lund, N. J., & MacKay, D. M. (1983). Sleep and the McCollough effect. *Vision research, 23,* 903–906 *and* Maguire, M. S., & Byth, W. (1998). The McCollough effect across the menstrual cycle. *Perception & Psychophysics, 60,* 221–226.

6. Bedford, F. L. (1997). False categories in cognition: The Not-the-Liver fallacy. *Cognition, 64,* 231-248 *and* Bedford reference in note 4, Chapter 1.

7. Buckland, R. (1969). *A Pocket Guide to the Supernatural.* Ace.
*Background image for Tree of Learning (Figure 12-2, facing end of chapter) is Tree of Life by Gustav Klimt, 1909.

Chapter 13

1. Wilson, B. A., Cockburn, J. A., & Halligan, P. (1987). Development of a behavioral test of visuospatial neglect. *Archives of Physical Medicine and Rehabilitation, 68,* 98-102.

2. Keltikangas-Järvinen, L, Pullmann, H., Pulkki-Råback, L. Alatupa, S., Lipsanen, J., Airla, N., & Lehtimäki, T. (2008) Dopamine Receptor D2 Polymorphism Moderates the Effect of Parental Education on Adolescents' School Performance, *Mind, Brain, and Education, 2,* 104-110.

3. The quote is from Theodore Zeldin, British sociologist.
Disclaimer: Nothing in the chapter or elsewhere is intended to be medical advice. Medical advice can be obtained only from your physician. Consult your physician if you feel you have a medical issue.

Acknowledgements

I would like to thank my students for the three E's: enthusiasm, energy, and (willingness to) engage in dialogue about anything. Maybe a fourth E, experiments, which could not have been done without them. They include: Anna Aper, Simone Chambliss, Tony Golembiewski, Erin Harvey, Birgitta (Gittan) Mansson, Whitney McNeil, Melissa Panagos, Sabrina Peterson, Karen Reinke, and Benjamin Travis. Likewise to students in classes, too numerous to mention, who were inspiring, resourceful, and who send me updates and questions from time to time.

Thanks also to Linda Falcão for editorial comments, Laura P. Garcia for tireless assistance with some of the illustrations, Whitney McNeil for a cartoon collaboration, Julia Britt for Swedish words (lost to editing), and Sabrina Peterson and Tony G for photos of themselves.

Finally, a special kind of thanks to those who struck me as supportive of the idea for this project from very early on (when others were not!) In no particular order, they are:

David Premack, Henry Gleitman, Lila Gletiman, Sue Spector, Paul Bloom, and Morris Moscovitch.

Munchas (not a typo) gracias,

Felice
Felice Bedford
bedford@u.arizona.edu
www.u.arizona.edu/~bedford

———

About the Author

Felice Bedford is on the faculty at the University of Arizona. She grew up in Brooklyn, New York and received a Ph.D. in Psychology from the University of Pennsylvania. She divides her time between Tucson, Arizona and a leafy suburb of Philadelphia. Besides the topics herein, collies, genealogy, genetics, medicine, and Korean dramas are her current passions.

Index

Author Index

www.ingramcontent.com/pod-product-compliance
Lightning Source LLC
Chambersburg PA
CBHW050450270326
41927CB00009B/1680